The Mechanized Battlefield
A Tactical Analysis

Pergamon Titles of Related Interest

RUSI/BRASSEY'S DEFENCE YEARBOOK 1985
RUSI/BRASSEY'S DEFENCE YEARBOOK 1986
Simpkin RACE TO THE SWIFT
Simpkin ANTITANK
Simpkin MECHANIZED INFANTRY
Simpkin TANK WARFARE
Simpkin HUMAN FACTORS IN MECHANIZED WARFARE
Simpkin RED ARMOUR

The Mechanized Battlefield
A Tactical Analysis

Edited by Lieutenant-Colonel J.A. English
Major J. Addicott
Major P.J. Kramers

PERGAMON-BRASSEY'S
International Defense Publishers

Washington New York Oxford London Toronto Sydney Frankfurt

Pergamon Press Offices:

U.S.A.	Pergamon-Brassey's International Defense Publishers, 1340 Old Chain Bridge Road, McLean, Virginia, 22101, U.S.A.
	Pergamon Press Inc., Maxwell House, Fairview Park, Elmsford, New York 10523, U.S.A.
U.K.	Pergamon Press Ltd., Headington Hill Hall, Oxford OX3 0BW, England
CANADA	Pergamon Press Canada Ltd., Suite 104, 150 Consumers Road, Willowdale, Ontario M2J 1P9, Canada
AUSTRALIA	Pergamon Press (Aust.) Pty. Ltd., P.O. Box 544, Potts Point, NSW 2011, Australia
FEDERAL REPUBLIC OF GERMANY	Pergamon Press GmbH, Hammerweg 6, D-6242 Kronberg-Taunus, Federal Republic of Germany
BRAZIL	Pergamon Editora Ltda., Rua Eça de Queiros, 346, CEP 04011, São Paulo, Brazil
JAPAN	Pergamon Press Ltd., 8th Floor, Matsuoka Central Building, 1-7-1 Nishishinjuku, Shinjuku, Tokyo 160, Japan
PEOPLE'S REPUBLIC OF CHINA	Pergamon Press, Qianmen Hotel, Beijing, People's Republic of China

---

Library of Congress Cataloging in Publication Data
Main entry under title:

The Mechanized Battlefield.

Proceedings of the First Annual Officer Development Study Session held at Canadian Forces Base Gagetown, Mar. 7-9, 1983.
Bibliography: p.
Includes index.
1. Tactics--Congresses. 2. Combat--Congresses. 3. Military art and science--History--20th century--Congresses. I. English, John A. (John Alan) II. Officer Development Study Session (1st : 1983 : Gagetown, N.B.)
U165.M46 1986 355.4′2 85-28431
ISBN 0-08-025405-5

Printed in the United States of America

"The object of all military training is to prepare the soldier for the next war, for it is the only possible war in which he can fight...."

---Major General J.F.C. Fuller,
CB, CBE, DSO

TABLE OF CONTENTS

TABLE OF CONTENTS

FOREWORD

The Combat Training Centre's First Annual Officer Development Study Session was designed to look at tomorrow, to how we should be preparing ourselves for the next war. The specific aim of the session was to examine, in both theoretical and practical terms, the tactical deployment of basic fighting units on the modern battlefield. To accomplish this, several prominent and eminently qualified military theorists and historians were invited to make appropriate presentations. Complementary briefings by active service veterans of the British Army, the United States Army, and the United States Marine Corps, as well as by Canadian army staff officers, served to round out this rather ambitious programme.

The three-day Study Session was deemed by most participants to be an unqualified success. The free exchange of ideas and information that occurred and the resultant discussion and tactical dialogue that developed proved remarkably stimulating. It was indeed a most thought-provoking forum, and, as any serious student of military affairs can attest, frank and candid professional military expression has more often than not led to better ways of doing things. At the Combat Training Centre, where the enhancement of tactical skills remains a day-to-day concern, this is particularly relevant. For officers of an army that last saw action in the Korean War, of course, the Study Session also provided a unique opportunity to profit from the combat experience of others. It is to be hoped that through the pages of these proceedings those not fortunate enough to have attended the symposium can share in its benefits.

Oromocto, New Brunswick
March, 1984

Brigadier-General J.J. Morneault,CD
Commander, Combat Training Centre
Canadian Forces Base Gagetown

PREFACE

This book derives principally from the First Annual Officer Development Study Session held at Canadian Forces Base Gagetown under the authority of the Commander of the Combat Training Centre. The advantage in having a permanent record of such proceedings is self-evident, as the occasion broached a number of military theories and tactical concepts that require more than just passing reflection. Two additional presentations made independently by Brigadier Richard Simpkin to military audiences in Ottawa have, for their appositeness, also been included as separate chapters. In all instances the innate strength of the printed word further serves to counter-balance a natural human tendency to recall only what one wishes to recall.

The purpose behind the American-Commonwealth gathering that took place at Canadian Forces Base Gagetown 7 - 9 March, 1983, was to stimulate thinking in the often neglected area of low-level tactics. A serious attempt was made to focus on operations below battle group (or battalion-size task force) level on the mechanized battlefield of the 1990s. The Study Session was designed, nonetheless, to consider the theoretical as well as practical dimensions of this particular sphere. The battle experience of combat veterans was accordingly juxtaposed with military theory, the foundations of which either rested in war game results or historical analyses.

In this volume, space has precluded the inclusion of formal question and answer sections. Those exchanges that were considered most pertinent, however, have been inserted at the end of each chapter under editors' notes. As for chapter order generally, neither symposium sequence of presentation and discussion, nor the unity of certain of the briefings delivered, determined how we structured our book. We have sought only to produce, if not a totally comprehensive account, at least an ordered and coherent one. We do hope that it will promote and enhance tactical thinking and understanding accordingly.

Without the full support of the Commander of the Combat Training Centre and the administrative staff of Canadian Forces Base Gagetown it would not have been possible to organize the Study Session from which this book stems. Additional thanks are also due to the Commander Mobile Command, the Army Doctrine and Tactics Board, and the Directorates of Land Combat Development and Land Operational Research, National Defence Headquarters. It must be understood, of course, that none of the views expressed in this book are necessarily officially representative of the above agencies or the Department of National Defence at large.

Within the Province of New Brunswick, special indebtedness is acknowledged to both Professor Dominick Graham, MC, of the Centre for Conflict Studies, University of New Brunswick, and Colonel James Turnbull, OMM, CD, of the United Services Institute of Saint John. We remain grateful, as well, to Majors Rupert Hartley, Paul Howe, and Mike Winter -- all of Tactics Division -- for the generous assistance they provided in reviewing first and last drafts; Sergeant Granger of the Tactics Intelligence Section also provided useful technical assistance. To our typists, Mrs. Leah Sabourin and Miss Sherry LeClair, who had the awesome tasks of transcribing tapes and preparing final camera-ready manuscript, we remain especially appreciative. To Major Dave Bianco and his Training Aids Section we can only offer our most sincere thanks for their unstinting and dedicated support.

Perhaps most important of all, we wish to express our sincere appreciation to our contributors, who, in most cases, travelled long distances and endured a certain amount of personal inconvenience and discomfort to be with us. Our proceedings greatly benefited from their candour and good comradeship, and we trust we have done them justice in the edited chapters that appear herein. To other participants we thank them for their questions and comments. Finally, we would like to acknowledge the vitally important and extensive editorial assistance given to us by our publisher, Mr. Don Crawley of Pergamon Press.

Tactics Division
Combat Training Centre
March, 1984

J.A.E.
J.A.
P.J.K.

LIST OF FIGURES

The Mechanized Battlefield
A Tactical Analysis

CHAPTER ONE: INTRODUCTION

by
Lieutenant-Colonel J. A. English, CD
Chief of Tactics
Combat Training Centre

It has always been most difficult to discern, with any degree of certainty, the form the battlefield of the future will take. More often than not, military professionals have been either technically or tactically surprised by the wars they have been compelled to wage. Yet, looking back on them, there is nothing particularly mysterious about them (as certain prescient individuals understood all along). According to General Sir Richard Gale, writing in 1955, two underlying causes of which we must always be aware serve to blind professional military foresight. In the first instance, we tend to form ideas too closely related to our last experience; secondly, we incline in peace to form too rigid ideas, not only in relation to our own tactics, but on those of any potential enemy. We attribute to the enemy a tactical method that is based on what he did in the last war and not on what he could do in the next.[1]

At the same time, military history does provide certain indicators and sign-posts that point toward the future. If history does not actually repeat itself, it can be said to paraphrase itself from time to time. There is really no reason, therefore, to have to go on re-learning all old lessons. Certainly, the Soviet armed forces are convinced that military history is relevant to them. Moreover, they consistently use historical military experience to solve contemporary military problems.[2] I must admit to being very much struck by the following passage from Viktor Suvorov's latest book, entitled *Inside the Soviet Army*:

> When I lecture to Western officers on tactics in the Soviet Army, I often close my talk by putting a question to them -- always the same one -- in order to be sure that they have understood me correctly. The question is trivial and elementary. Three Soviet motor-rifle companies are on the move in the same sector. The first has come under murderous fire and its attack has crumbled, the second is advancing slowly, with heavy losses, the third has suffered an enemy counter-attack and, having lost all of its command personnel, is retreating. The commander of the regiment to which these companies belong has three tank companies and three artillery

> batteries in reserve. Try and guess, I say, how this regimental commander uses his reserves to support his three companies. "You are to guess", I say, "what steps a Soviet regimental commander would take, not a Western one but a Soviet, a Soviet, a Soviet one."
>
> I have never yet received the correct reply.
>
> Yet in this situation there is only one possible answer. From the platoon level to that of the Supreme Commander all would agree that there is only one possible decision: all three tank companies and all three artillery batteries must be used to strengthen the company which is moving ahead, however slowly. The others, which are suffering losses, certainly do not qualify for help. If the regimental commander, in a state of drunkeness or from sheer stupidity, were to make any other decision he would, of course, be immediately relieved of his command, reduced to the ranks and sent to pay for his mistake with his own blood, in a penal battalion.[3]

I expect that Professor Vigor, who has just published his new book Soviet Blitzkrieg Theory, will go into these tactics in greater detail when he discusses the Soviet theory of echeloning. I would say, however, that such tactics are not new: Liddell Hart recommended them years ago as a basic infantry attack tactic and the Germans used the same approach in their central concept of the Schwerpunkt, or "thrust point" of principal effort that was constantly altered to fit the circumstances of seeking a line of least resistance through enemy defences. It is the "expanding torrent", the reinforcement of success rather than of failure, and I wholeheartedly agree with Suvorov that it is not a weakness but a strength. It also has some serious implications for us, for we will have to respond with movement, hopefully, not to reinforce, but to spoil or destroy; counter-attacking to kill rather than to recapture and hold lost ground.

Having mentioned the requirement for battlefield movement and mobility, which subjects both Brigadier Simpkin and Major Charles Messenger will address, I would now like to broach the subject of digging. We know from German army and other troop trials that prone infantry unprotected in the open will likely suffer 100 per cent casualties if subjected to artillery and mortar fire of the intensity prescribed in Soviet doctrine.[4] Red Army high density rate of artillery fire would probably be around 40 rounds on one hectare (a 100 metre by 100 metre square) per minute.[5] I am, as well, given to understand by our war games people that infantry combat vehicles deployed in defensive positions also might not survive such weight of fire. Fortunately, a recent British trial revealed that what we call the "DePuy" parapet trench and the two-man battle trench with overhead protection both stand up reasonably well. It is hoped that Liddell Hart's argument (based on his observations of Sherman's Civil War campaigns) that " trenches and mobility are not irreconcilable"[6] will also continue to stand up.

The requirement for infantry to move rapidly and securely from one dug-in position to another is, of course, of primary concern, and one that the proposed Canadian infantry combat vehicle (CICV) can doubtless meet. The question of how we employ its cannon, however, is a little bit more complicated. Obviously it will give our infantry the capability to kill enemy infantry fighting vehicles (IFVs) at 1000, possibly 2000 metres. This will theoretically force enemy infantry to dismount at an earlier stage, leaving them vulnerable for a longer period to our direct and indirect fire, which, in recent wars, has caused upwards of 50 percent casualties (though some people would say more, on the Eastern Front it probably ran to about this). It will also allow our other anti-armour weapons to concentrate on killing tanks. Clearly, it would be a waste to leave these vehicles with their respective 14-odd kinetic energy weapons sitting in a rear harbour, as was, and continues to be, practised with the armoured personnel carrier (APC). Yet to place them forward in sniping positions might expose them to the fire of enemy tanks and attack helicopters, in which circumstance our infantry could end up back on their feet as in 1915, unable to speed from one dug-in position to another in relative security. Deploying CICVs to flanks or even in depth supporting a reverse slope position might be a better answer. I am confident, of course, that Major-General J.F.C. Fuller was quite correct in stating that:

> Given any weapon, however novel, and the circumstances in which it is likely to be used, and the characteristics of the soldiers who will use it, it should always be possible to work out for it a sound tactics, irrespective of the fact that it has never been fought with.[7]

The foregoing problem leads naturally into one of the fundamental questions of modern warfare, and that is: the proper use of infantry. Always the most vulnerable man on the battlefield, the infantryman continues to be gravely threatened by the armoured vehicle, and the threat from tanks is so great that he cannot possibly survive without an anti-tank capability; more than this, he also needs to be innoculated against that well-documented and well-recognized psychological affliction called "tank fright". According to the late Colonel John Weeks, the specific tasks of infantry, within a general role of providing a battlefield framework, are:

a. To hold ground against enemy armour and infantry attacks and provide a firm pivot for counter-attacks or other manoeuvres;

b. To dominate and control the close country;

c. To close with the enemy and clear him from his defensive positions (winkle him out, so to speak, and get him to sign the surrender document); and,

d. To carry out observation, reconnaissance, and provide early warning.[8]

To General George Patton the relationship between infantry and armour was quite clear. In an infantry division, the purpose of tanks was to use their guns to get the infantry forward, while in an armoured division the function of infantry was to break the tanks loose to ranges where they could use their machine guns on enemy infantrymen.[9] In North Africa it was common practice for German tanks to turn tail and lead superior numbers of British tanks onto well-concealed anti-tank screens. In these actions, British tank crews, having had their tanks actually shot out from under them by an invisible 50mm anti-tank gun, often ascribed their plight to the more active and visible German tank. It took them a long time to realize that they were being knocked out, not by the tank, but by those handy low silhouette guns. The remarks of General Frederick J. Kroesen, former commander of NATO's Central Army Group, are especially pertinent in this regard:

> We constantly hear someone saying, "The best anti-tank weapon is another tank." While the logic of that statement is difficult to dispute, as tanks combine the essentials of firepower, mobility and protection, its accuracy has not always been accepted. In the aftermath of Operation Crusader in 1941, General Gott, Commander of the British 7th Armoured Division, wrote, "The German will not commit himself to a tank versus tank battle. In every phase of battle he coordinates the action of his tanks with his anti-tank guns, field artillery and infantry and will not be drawn from this policy." Another veteran of the 7th, Brigadier John Strawson, stated: "Rommel's Panzers were quite clear that, whereas tanks dealt primarily with the enemy's infantry and soft-vehicles, the destruction of tanks was mainly the job of anti-tank guns."[10]

Soviet advisors to the Egyptians in the 1973 Yom Kippur War evidently took these fundamentals into account. In that war, Egyptian tanks were purposefully held back as bait for Israeli armour; and we all know what happened to the 190th Armoured Brigade in its subsequent encounter with Egyptian infantry armed with anti-tank guided missiles (ATGM) and RPG 7s. In the NATO scenario, of course, there is reason to believe that if one places tanks forward in any battle group[11] defensive deployment, the risk of their destruction by anti-tank attack helicopters is greatly increased; in depth, on the other hand, where the latter are themselves vulnerable, tanks are able to deliver over familiar ground of their own choosing a good offensive punch. Given that properly sited and adequately armed modern infantry now appear to possess a reasonable capability for taking on either of the aforementioned weapons systems, future tactical engagements could well bear a resemblance to the old game we used to play as children: rocks break scissors, scissors cut paper, and paper covers rocks.

This makes reasonable sense, moreover, since dismounted infantry have traditionally not wanted tanks around them (unless they were in dire straits) because tanks draw fire. Then, too, whereas tanks normally look for longer fields of fire in open country within which they can jockey, the infantryman looks to closer country within which he can take cover and fight at close quarters from prepared positions not subjected to long-range direct fire and observed indirect fire. Some even argue that infantry should hold fire until the enemy are within a few hundred metres and then stop them dead. Fire opened too soon can miss, and a miss discloses one's position. I am simply amazed at the number of Second World War after-action reports that stress the vital importance of holding one's fire until almost the last minute in order to be sure of killing an enemy (even the feared Tiger tank).[12] Might one not suggest, then, that if Soviet infantrymen wish to motor to their deaths by IFV load, we oblige them? The old

German defensive principles of maximum depth, invisibility, and readiness for action would obviously apply in such case.

Having ventured that, I would now like to raise the subject of engagement ranges in North-West Europe. The Soviets consider 50 per cent of Germany passable to tanks and, according to Chris Donnelly, accept West German target sighting distances. They thus anticipate that in anti-tank engagements 60 per cent of targets are likely to be acquired at no more than one kilometre.[13] Modern defence, it would seem, thus depends as much on short-range fire as on long. To punctuate this point I shall once more quote General Kroesen:

> We cannot hit what we cannot see and 14 hours of darkness in (Germany) mid-winter, snow, rain and the many days throughout the year when fog lasts until noon or even all day are limitations that today's weaponry cannot readily overcome. The same is true of our opponent's weapons. Those realities and the availability of tactical smoke generating devices in abundance lead me to believe that the next war will be won or lost at the 300m range just as in the past.[14]

As for attacking over this distance with infantry, I believe we have much to learn from Major Collett and Major Livingston, both of whom have personal experience in this area. It is perhaps worth mentioning, however, that in Current Reports from Overseas dated 10 January, 1945, a rifle company commander wrote:

> It has been proved to my battalion that the Brens are the guts of the platoon -- the framework, as it were, on which the rest of the platoon is fashioned. Consequently, to get the best value out of the Brens we now keep four Brens in each rifle platoon, forming two Bren sections each of two Bren groups.[15]

I find this particularly interesting, since it was the strong recommendation of the Second Battalion, The Parachute Regiment, in its post-action report from the Falklands War that the rifle section be established and organized with two GPMGs, belt-fed. There, they conducted fire and manoeuvre at section level between two fire teams, finding it a far more satisfactory and effective arrangement than between a rifle and gun group. I might add in the same breath that while we have already designed and accepted a new section structure as a result of our Canadian army infantry study, we remain especially interested in the rationale underlying the new Marine squad structure (the old organization having served the Corps so well for so many years, some 40-odd I believe). Significantly, Major-General Franz Uhle-Wettler, former commander of 5 Panzer Division, recently recommended for his proposed light infantry division structure large infantry sections of roughly thirteen, "like those of the United States Marine Corps". Interestingly, Major F.O. Miksche as early as 1941 recommended a similar 16-man section, for which structure lots of good arguments, I am quite sure, can still be put forth today.[16]

On that note I shall conclude, albeit with a caution that though the purpose of this symposium is to examine lower level tactics on the modern battlefield, the tactical deployment of smaller units often depends so heavily upon the larger picture that it might be difficult to maintain such focus with the constancy desired. Regrettably, too, the air factor will not receive the attention that its likely prominence on any future battlefield would seem to warrant. It is hoped, nonetheless, that all participants will derive substantial professional benefit from the free exchange of ideas that ensues, whatever the degree of focus attained.

I now recommend that we watch and shoot, but only in that relaxed and informal manner that tactical brainstorming calls for. We have here today amongst us, some of the finest military theorists in the Western world and some of the most experienced combat officers, several of whom -- our allies -- have actually seen action. We must make the most of this, and I therefore direct my final comment to the young officers of the Combat Training Centre, and again it is to quote J.F.C. Fuller, who in former times addressed one of his military classes as follows:

> You have come here to acquire knowledge, to evolve it and to fit it to men you will one day either command or administer. Knowledge is not only acquired in the lecture room, but also in the Mess and in your own private studies. Nothing clarifies knowledge like a free exchange of ideas; consequently, because I happen to be a Colonel and you a Captain or a Major, do not for a moment imagine that rank is a bar to free speech. If you disagree with me or the views of any of your instructors, openly state your disagreement, for we are all students, and the man who cannot change his opinons has mineralized his intellect -- he is a walking stone.[17]

So I say to all young officers, participate.

Notes

1 General Sir Richard N. Gale, "Infantry in Modern Battle", Canadian Army Journal, No 1 (January, 1955), 52-53. He wrote as Commander-in-Chief, British Army of the Rhine.

2 Colonel T. N. DuPuy, "History and Modern Battle", Army (November, 1982), 19 and 22.

3 Viktor Suvorov, Inside the Soviet Army (New York: MacMillan, 1982), 171.

4 Lieutenant-Colonel Robert G. Chaudrue, "Requiem for the Infantry", Infantry, No 3 (May-June, 1978), 29.

5 C. N. Donnelly, "The Wind of Change in Soviet Artillery", International Defense Review, No 6 (1982), 739 and 741.

6 B. H. Liddell Hart, Sherman (London: Ernest Benn, 1930), 136.

7 Major-General J.F.C. Fuller, Memoirs of an Unconventional Soldier (London: Ivor Nicholson and Watson, 1936).

8 Colonel John Weeks, "The Modern Infantryman", Military Technology and Economics, Issue 9 (May-June, 1979), 23 - 24; and same issue, DMO Miller, "The Infantry Vehicle: An Assessment", 29 and 38.

9 General George S. Patton, Jr., War As I Knew It (New York: Pyramid), 307.

10 General Frederick J. Kroesen, "The Ultimate Weapon of War", RUSI Journal (December, 1980), 63.

11 An all-arms grouping based on a battalion-size unit and comprizing a number of combat teams based on company-size sub-units.

12 See, for example, "A Few 'Tips From the Front'" Canadian Army Training Memorandum, No 29 (August, 1943), 20; "What an Infantry Subaltern Really Is", Canadian Army Training Memorandum, No 37 (April, 1944), 42; Current Reports From Overseas, No 15 (September 1943, 8-9; Current Reports From Overseas, No 36 (May, 1944), 13; and Current Reports From Overseas, No 77 (February, 1945, 9.

13 Anti-tank engagements are expected in the following ratios: 40 per cent at ranges under 500 metres, 20 per cent at 500 - 1000 metres, 25 per cent at 1000 - 2000 metres. They quote West German figures that terrain and climate will allow on average for 70 per cent of tank targets to be observed at 2000 metres. C.N. Donnelly, "Soviet Tactics for Overcoming NATO Anti-Tank Defenses", International Defense Review, No 7 (1979), 1101.

14 Kroesen, 64.

15 "Battle Impressions of a Rifle Company Commander", Current Reports from Overseas, No. 71 (1945), 5. Reference is to the Bren light machine gun, which was replaced by the general purpose machine gun (GPMG).

16 Udo Philipp, "NATO Strategy Under Discussion in Bonn", International Defense Review, No 9 (1980), 1370; and Major F. O. Miksche, Blitzkrieg (London: Faber and Faber, 1941), 19, 133-135, 189.

17 Fuller, Memoirs, 419 - 420.

CHAPTER TWO: THE WARSAW PACT THREAT

A CANADIAN ANALYSIS

by
Lieutenant-Colonel I. D. Barnes, CD, and
Major N. B. Jeffries, CD, Directorate of Land
Combat Development (DLCD), National Defence
Headquarters, Ottawa, Ontario

> The Combat Training Centre's First Annual Officer Development Study Session opened with a major DLCD presentation that provided overviews of the Canadian land forces combat development[1] process and the "Corps 86" concept, the latter essentially the result of the former and the conceptual umbrella under which the Canadian army will prepare for the conduct of land warfare in the 1986-1995 timeframe. Suffice to say, the development of the operational concept for a Canadian corps operating against a sophisticated enemy, such as the Warsaw Pact, in a high-intensity combat environment during the aforementioned timeframe was based upon a number of factors. Strategic assessment defined the threat, political and executive direction outlined the level and capabilities the Canadian Forces were to develop to counter same, technology established the parameters of weapons capabilities, and threat analysis determined what a Canadian corps must do to prevail in a high-intensity battle against a sophisticated enemy. What follows in capsulated form, is an excerpt from the DLCD presentation that specifically concerns the nature of the threat.

As the Warsaw Pact threat figures prominently in the deliberations of this study session, we will review for you its salient points as viewed from a Canadian perspective. In essence, the Canadian army threat analysis considered the broad organizational and doctrinal characteristics of a Warsaw Pact front (army group) organization. The aim of the threat analysis was to assess enemy strengths and weaknesses to determine what could be done to overcome and exploit them.

The major organizational and doctrinal characteristics of the Warsaw Pact front forces were considered to be: echelonment of forces; a massive firepower capability (provided by nuclear, biological and chemical (NBC) weapons and an imposing weight of armour and artillery); and, finally, the doctrine of mobile operations. We will proceed to examine each one of these characteristics in detail, starting with echelonment of forces.

Warsaw Pact forces from motor rifle regiment to front will be deployed on the battlefield into echelons. The enemy will use its large numbers and its echelonment of forces in both offensive and defensive operations to engage our forward forces, while at the same time retaining a large portion of its forces uncommitted. In the defence, Warsaw Pact uncommitted forces form the framework of the depth and reserve. In offensive operations, as Warsaw Pact first or forward echelon forces are reduced to an ineffective level, extend their resources, or establish a breach in our forward defences, they will be continuously replaced with fresh second echelon or follow-on forces to maintain the momentum of the advance. The tempo of the Warsaw Pact offensive through echelonment of its forces is such that there would be little time for a defender to relieve or reinforce between the first and second echelon assaults. One method of countering this effect is to build durability into our forces. This would be achieved by including sufficient units to permit deployment in depth and the constitution of a reserve. Greater durability could also be catered for through development of larger units and formations.

An additional method of countering Soviet echelonment of forces is to attempt, in the defence, to disrupt and delay the progress of the second echelon, and, in the offence, to prevent the second echelon depth and reserve forces from redeploying. In the defence at corps level, disruption and delay of Warsaw Pact second echelon forces should allow our forward divisions to stop and destroy the assaulting first echelon; in other words, the momentum of the attack would be reduced by preventing the second echelon from influencing the battle on the forward edge of the battle area (FEBA). In the offence, engagement of second echelon depth forces and reserves should give friendly assault forces time to break through enemy forward defences and strike deep into rear areas. To separate enemy first and second echelon forces, in both offensive and defensive operations, is thus the major deduction drawn from analysis of this organizational characteristic.

The Warsaw Pact's massive firepower capability is, of course, reflected in its nuclear and chemical arsenal, sheer numbers of artillery weapons and tanks, and their planned tactical use. It constitutes a formidable threat. Warsaw Pact forces are, in addition, well-trained and equipped for both the offensive and defensive employment of nuclear and chemical weapons, which they are prepared to use to ensure success or avert failure. Once release is given for the employment of chemical weapons, the Warsaw Pact will treat them as but another item in their conventional arsenal. While Canadian policy at this time precludes the initial use of nuclear and chemical weapons, it was recognized that Canadian troops could be required to operate in a nuclear and/or chemical environment, The need to provide for a defensive capability to include detection, protection, and decontamination of personnel, weapons, and equipment was accordingly identified. It was also considered prudent to retain a planning capability against the possibility that political authority might at some point authorize the employment of nuclear or chemical weapons, either directly or indirectly, as a retaliatory measure.

In Warsaw Pact doctrine, the tank remains the major striking force. It is located at all levels from motor rifle regiment through to combined arms and tank armies, as well as in reconnaissance units. To counter this preponderance of armour, a complete range of anti-armour weapons systems is required at all levels, from company through to corps. Attrition of enemy armour must commence as early as possible within weapons system capabilities so that the enemy's numerical superiority is reduced before he can engage in the direct fire battle. An offensive shock action capability, with which to alter the course of the battle in friendly force favour, is also required. As the tank remains the most flexible and effective anti-tank weapon available, a strong mobile reserve based upon tanks should be included in defensive deployments. Possible tasks of such a reserve would be to block an enemy breakthrough, to restore the FEBA, to mop-up remnants of first echelons, or to secure a firm base from which to launch offensive operations. In the offence, mobile forces based on tanks would also be required to push deep into enemy rear areas or conduct pursuit operations once a breakthrough has been achieved. The main deduction drawn from this analysis was that high intensity conflict in the 1986-1995 period will be, at least in the initial phases, an anti-armour battle. Specifically, therefore, an armour-heavy formation should be included at corps level.

Soviet T-72
Fig. 2-1

Turning now to Warsaw Pact artillery, Canadian forces can expect to be confronted by an impressive array of mortars, guns, rockets, and missiles, including nuclear and chemical munitions. Dependent upon enemy force concentrations, friendly artillery could be outnumbered by as much as five to one. To survive the threat posed by Warsaw Pact artillery and its complementary target acquisition capability, it was concluded that regardless of the type of operation it would be necessary to fight from well-prepared positions. Furthermore, to ensure the survivability of friendly artillery as well as provide Canadian troops and equipment with adequate protection and relief, a requirement for a comprehensive counter-bombardment program was identified. The capability to detect enemy artillery before as well as during firing would have to be an integral part of such a program.

The Soviet doctrine of mobile operations is reflected in the capability of the Warsaw Pact to conduct large-scale, fast-moving operations both on land and in the air. The majority of enemy armoured vehicles are tracked and can swim, snorkel, or deep-ford. The enemy also has considerable engineering resources -- including bridging equipment -- deployed well forward to assist in overcoming any obstacles. Using both fixed and rotary wing resources, the enemy is additionally able to penetrate deep into NATO rear areas to secure strategic and tactical objectives, such as bridges, or to attack lucrative targets such as nuclear, logistic, and headquarters facilities. It was deduced that to counter this threat Warsaw Pact ground forces must be slowed in their advance in order to expose them to fire for as long as possible. To achieve this, it was concluded that defensive operations must be based on an obstacle or series of obstacles. Sheer numbers, however, still give Warsaw Pact forces the strength and capability to break through forward defences at selected places; any defensive concept, therefore, must include forces in depth to block penetrations. Mobile forces held in reserve will also be required to close with and destroy Warsaw Pact forces once they are stopped. Finally, it was concluded that to counter airborne landings in the rear area, dedicated forces would be required.

In short, although NBC and artillery capabilities are major aspects of the threat, the focus of the enemy's organizational and doctrinal thinking remains centred on his armour. He will conduct intense, sustained offensive operations involving concentrations of successive echelons of armoured vehicles capable of converging onto narrow breakthrough frontages where favourable combat ratios of 4:1 to 6:1 will be achieved. In a "worst case" scenario, with a Soviet tank division attacking on a breakthrough sector, first echelon strength assaulting on a one kilometre frontage would include 30 main battle tanks and 20 infantry fighting vehicles carrying 160 dismountable infantrymen. This assault would be supported by suppressive fire on a scale of 100 artillery tubes per kilometre of front, vehicle mounted air defence weapons, armoured reconnaissance and engineer vehicles, fixed and rotary wing tactical aircraft, and extensive electronic warfare jamming equipment. To defeat such an attack, or to win time by causing the enemy to replace non-effective units, will require the destruction of up to 65 per cent of the initial assaulting force. The battlefield environment will increase the dimensions of this challenge, as research shows that up to 70 per cent of armoured vehicles in the assault will not be detected by

Soviet BMP
Fig. 2-2

ground observers beyond a range of 2,000 metres. Based on the anticipated speed of the assault, there will thus be only seven minutes in which to reduce a force of 50 armoured vehicles by 65 per cent (an additional threat from dismounted infantry will also materialize if troop-carrying vehicles are permitted to reach reasonable dismount areas). All of this must be accomplished under the effects of neutralizing fire, obscuration, and the psychological impact of high intensity battle fought under the constant threat of chemical and tactical nuclear weapons.

The threat posed by this "worst case" situation in high intensity conflict in central Europe is the most demanding challenge to be faced by Canadian land forces. It is believed that forces structured and equipped to defeat this threat can, with the addition of specialized equipment and acclimatization to a particular environment, be effective against lesser threats in other scenarios.

SOVIET ECHELONMENT OF FORCES

by

Professor P. H. Vigor
Soviet Studies Research Centre
Royal Military Academy, Sandhurst
Camberley, Surrey

The hard work of our unit is spent in getting our material from Soviet sources, and since all this is done from Soviet sources, inevitably it is the Soviet viewpoint that I am giving you. It may therefore conflict quite considerably with your particular views on echeloning; but unless somebody can prove to me that I have misread the Russian, I shall continue obstinately to feel that what I am putting forward to you today is what I ought to be putting forward in my capacity as interpreter of the Soviet scene.

In the Soviet view, echeloning can be applied to defensive as well as to offensive operations, but since from the standpoint of any NATO force, Soviet offensive operations are likely to be of greater interest than Soviet defensive operations, I shall confine my remarks today to their missions of echeloning in the offence.

The Soviet understanding of the phrase and of its meaning is fairly precise which, in itself, is a considerable difference from the Western view. Take, for instance, the U.S. Glossary of Military Terms -- JPS 1, look under echeloning, and you will find there are four different meanings given. The Russian word echelon essentially has one meaning. Admittedly, it does have a meaning of a "train load", but that is not particularly pertinent to our subject today. For our purposes the meaning is precise and it begins by making a clear distinction between a second echelon and a reserve. In Soviet terminology a reserve is a unit or formation that the commander keeps back at his disposal for use during the battle for tasks that have not been pre-set. In Soviet practice, incidentally, the reserve is generally very small. Typical for an army group would be two divisions maximum in reserve; these would form a group that it has been decided to keep back from the initial stages of the battle, but the commander of the group will not know when the battle starts what it is he has got to do.

The second echelon, however, has always had very precise tasks; traditionally, its basic function in Soviet theory and practice was to take over from the first echelon of the attack when the latter had just about exhausted itself. In Soviet terms this generally meant incurring about forty per cent casualties. When this happened, the second echelon was committed immediately, and was expected thereby to keep up the pressure on the enemy. This was a very important part, and remains a very important part, of the theory. Keep up the pressure and the defenders have no opportunity to rest or to regroup; and if the air support is doing its job properly, the defenders are not going to get the opportunity to resupply and reinforce either. In the last war, what was demanded of Soviet army commanders, and of course Soviet group commanders, was that their first echelons should, within twenty-four hours, have so weakened the enemy first line of the defence that the second echelon would have no difficulty in moving through it. This meant in practice, and, of course, continues to mean, that in coping with a deeply echeloned defensive position two echelons are simply not enough, because the second echelon is going to wear itself out before it has got through. Certain of the German defensive lines in East Prussia, for instance, were something like 15 kilometre thick, with intervals of several kilometres between one set of defensive positions and the next set of defensive positions. The Soviets were, on these occasions, thus forced to adopt three echelons and, in one or two instances, even four.

As the European battlefield we are envisaging is not likely to have such echeloned defensive positions, I shall consequently limit my presentation to the deployment of two echelons only. If we agree, then, that the second echelon is the one that takes over the job of the first echelon when it is spent, we can illustrate the Soviet theory of echeloning by referring to the army group deployment shown in Fig. 2-3. I am using the term "army group" instead of the Russian word front because I am trying to keep your minds concentrated on the actual principles of the theory. We are trying to be as abstract as we can.

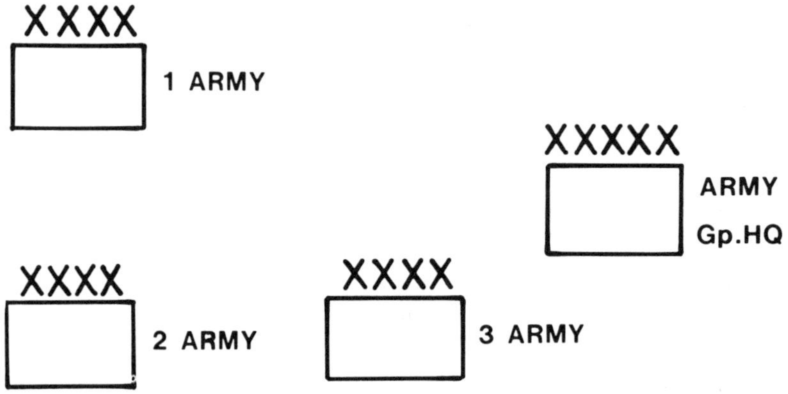

Echeloning to Army Level
Fig. 2-3

Fig. 2-3shows an attacking army group headquarters that has decided to echelon its armies in the fashion I have mentioned. It has made Third Army the second echelon of Second Army, and First Army has no second echelon at all. This is for the standard Soviet reason that what you support is the line of advance that you anticipate is going to have the greatest likelihood of success. The Russians are great believers in reinforcing success, and I shall elaborate a bit more on this later. I would stress, nonetheless, that if it had been the decision of the army group commander that it was First Army axis that was going to be the more favourable one, then, of course, the Third Army would be deployed behind it. As things are, the role of the First Army is simply going to be to launch what you might call either a diversionary or a holding attack, while the real job is done by the Second and Third Armies.

Echeloning is not just done by army groups and armies, it flows through the entire spectrum from the very top to the very bottom of the Red Army. In the Great Patriotic War the Russians had groups of army groups and, therefore, had echelons of army groups. If we leave these and come down to slightly more basic levels, however, we can see in Fig. 2-4, not only that any army group will echelon its armies, but that the army will, in turn, echelon its divisions. It does not necessarily follow, of course, that the army commander will wish to echelon his divisions in the way that his superior officer has echeloned him. In this particular example, I have shown 22nd Division as the Second Army's second echelon. I have placed it this way because, in the situation I am envisaging, the Second Army commander is not certain whether 22nd Division's advance will go better than 21st Division's. The Second Army Commander would therefore tend to put the second echelon more or less geographically midway between in order to be able to switch support to whichever seems the most promising line of attack. Obviously this can only be done if the terrain permits. If, for instance, one is moving in very mountainous territory around defiles, one might like to echelon along the lines shown in Fig. 2-4, but it is really very improbable that such ground will allow one to do so. Please bear in mind that I am speaking theoretically here, and that the Russians are not such fools as to neglect the practical constraints which factors like terrain are likely to impose. One cannot echelon in a vacuum, whether echeloning at army, divisional, or regimental level.

So far I have spoken of echelons as if the use of a second echelon is always profitable and to be desired, and that there is no price to pay for using them. This is simply not so: there is a price to pay and it is often quite a heavy one. The price is that to deploy in more than one echelon substantially reduces the amount of firepower that an attacker can bring to bear at the start of the battle. If trying to surprise an enemy strategically (and the Russians at the present moment are terribly interested in the question of surprise), one would obviously like to have the maximum firepower that the situation confronting one would allow. Yet, having paid such a price, a second echelon enables one to keep up pressure on an enemy.

To turn for a moment to the matter of surprise, it is of no use just surprising an enemy and thinking that this is all one has to do. An enemy can be totally surprised, but if one does nothing else than just surprise him, the enemy will undoubtedly recover. At Messine's Ridge in the Great War, the Allies with their huge mining operations surprised the Germans no end, but General Sir Herbert Plumer's British force [2] was not able to exploit the surprise sufficiently effectively, most Soviet strategists attributing this to the fact that Britain is a small country and that small countries really should not fight wars because they cannot afford to lose many men. A normal Soviet view on war is that wars are won by losing men -- lots,

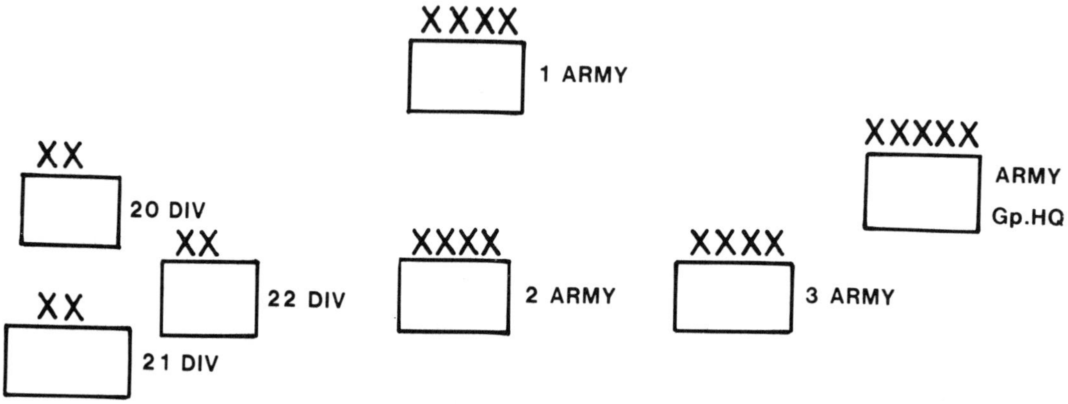

Echeloning to Division Level
Fig. 2-4

and lots, and lots, of men. Regardless of whether it is accomplished by manpower, however, or by speed and mobility (the modern Soviet tendency), surprise in order to be successful must see the elements of firepower and manoeuvre brought to bear right from the beginning.

To return now to echelonment, Fig. 2-5 is representative of a land mass that in its middle bears a slight resemblance to the European continent but which at its northern and southern extremities has certain differences. I have chosen such ground deliberately, because once again I want to keep this talk as theoretical as possible (which is also why I have not talked about red and blue forces). You will see that I have postulated that, up in the north, there is very difficult terrain; that the central sector is flat country like European Russian and the North German Plain; and that down in the south there is country which resembles the Ardennes. Such country, you will remember, is traversable by armour.

Fig. 2-5 shows how Army Group Centre has, in this particular instance, decided to deploy his subordinate armies. Since it is flat territory, he does not appear to be quite sure which axis is going to be the most profitable. He has deployed in two echelons because he believes the enemy will think that, because it is flat territory, that is the way the attack will come if it comes; he has, therefore, got to be prepared for a resistance. So he has deployed in two echelons and put them in their positions. Army Group North, faced with very difficult terrain, has decided to deploy in three echelons

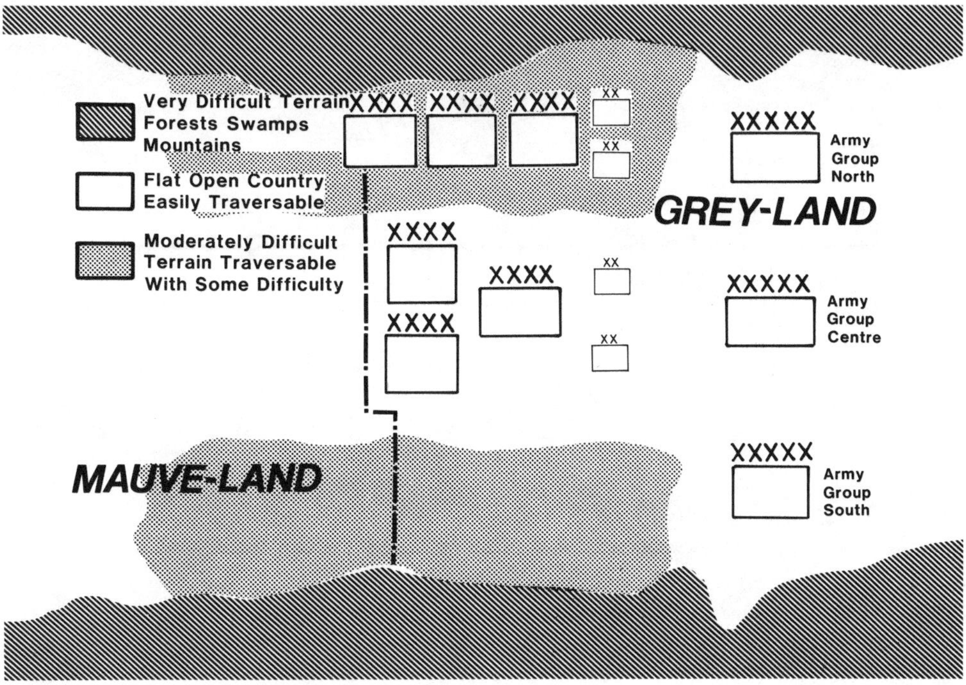

Comparative Echeloning
Fig. 2-5

with two divisions in reserve. Army Group South, following Hitler's example in the Ardennes in 1940, has gone in for one echelon, again with a small reserve. I would like to make it clear at this juncture that such deployment is not rigid teaching in the Soviet military schools. In other words, they do not say, if you are faced with really difficult country, you must deploy in three echelons. A problem of this nature in Soviet military institutions could call up various solutions, providing that basic principles are respected. That is the important point.

Another possible scenario is shown in Fig. 2-6. Greyland is thinking of attacking Mauveland, and has decided that it will gain maximum surprise with maximum opportunity for exploiting it by putting every army into the first echelon and having no second echelon at all (if you say that no army group commander would ever do this, I would refer you to the map at Fig. 2-7 which shows a Soviet army commander doing precisely that by deploying twelve armies in single echelon). In order to maximize the weight of blow, you will see that the reserve has been reduced from two divisions to one. I would reiterate, of course, that because the army group commander has deployed his armies in such fashion does not mean that his army commanders would deploy their subordinate divisions in single echelon. Normally they would be deployed in two. I could illustrate other possibilities that I have come across in Soviet writings. Obviously, in the nuclear age commanders must as well remember that too great a concentration of forces is liable to make a tempting nuclear target.

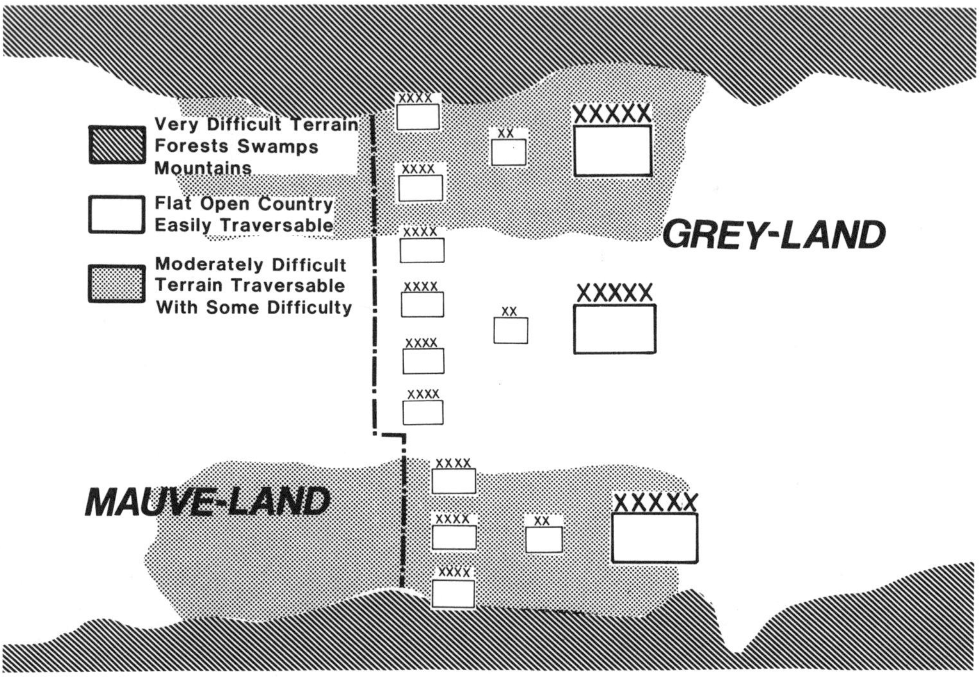

Single Echelon Deployment
Fig. 2-6

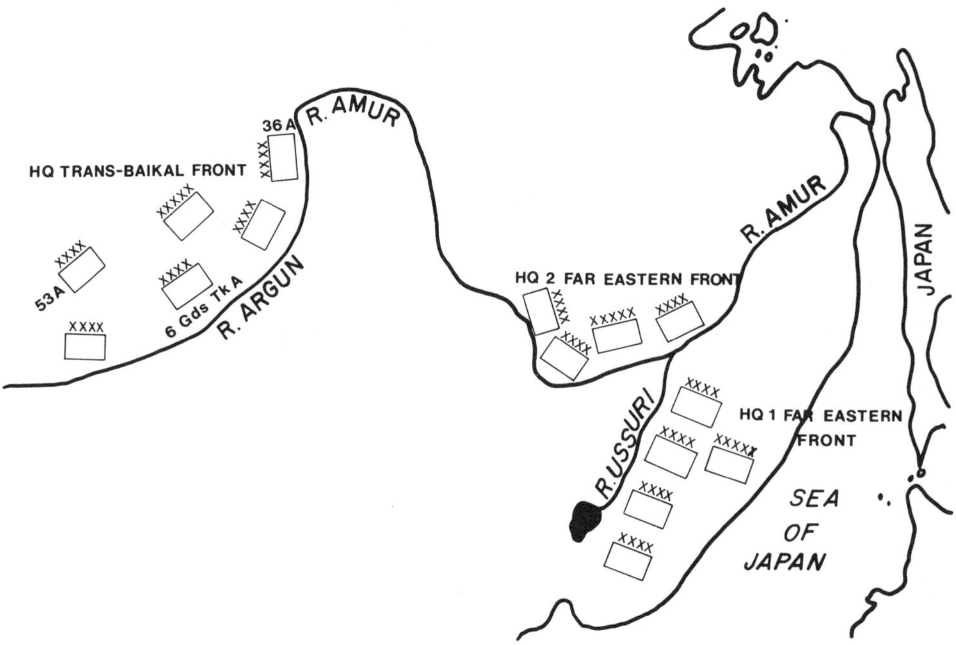

Echeloning of Soviet Forces in their
Surprise Offensive on the Japanese in Manchuria, 1945
Fig. 2-7

Fig. 2-7, in fact, shows the actual deployment of Marshal A.M. Vasilevskii's forces in Manchuria in August, 1945. It was his aim to achieve strategic surprise over the Japanese. He had three army groups, and, as indicated, he echeloned them all virtually in one echelon. Strictly speaking, it is a little incorrect to say that the army group on the left was deployed in a single echelon because 53rd Army at the beginning of operations was acting as a reserve, not really as a second echelon. Because Sixth Guards was a tank army, and because its job was to race ahead from the main body as fast as it could go -- over desert and the Khingan Mountains -- there was inevitably going to be an ever-widening gap between it and the rest of the army. The job of 53rd Army was to fill this gap. Vasilevskii was, of course, counting upon attaining surprise, which the sending of a tank army through the mountains accomplished.

There is one point I would like to bring out, if I may, and that concerns the army on the left, the nearest one towards the R in the River Amur. Just over the river, there is a main road and a main railway and a subsidiary river running east and west; also there, and commanding these features, lies the town of Hailar. Since it was on the only route from the west into Manchuria, it was obviously likely to be attacked; at least this is what the Japanese High Command thought, and they consequently had it heavily fortified. The Army Group Commander, Marshal R. Ya. Malinovskii, in fact, directed 36th Army to attack Hailar, though as it happened, he was not actually very interested in it. On the other hand, had he not attacked Hailar, the Japanese Command might have suspected that something was going not quite as they thought it ought to go. The interesting thing about the whole affair is that 36th Army was quite deliberately made to be the sort of army that Shakespeare's Falstaff took with him. It was composed of odds and sods, many people who had just come out of hospital, and lots of people who had never seen a battle in their lives. They were put under an absolutely first class army commander, who had done extremely well in Europe fighting the Germans, and were launched against Hailar in order to be massacred to keep the Japanese happy. They succeeded admirably. While the Japanese killed large numbers of irrelevant Russians and were terribly pleased with themselves, the main Soviet attack went in in another direction. The main centre of Japanese resistance was simply and quickly smashed to pieces.

Up to now I have spoken of the function of a second echelon as being exclusively that of replacing a first echelon which is unable to complete the breaching of an enemy defensive position. That was indeed the concept on which the whole process of echeloning was founded, at least so far as the Russians are concerned.[3] That has now changed, or at any rate is in the process of changing. Because modern defence is unlikely to be conducted behind great lines of massive fortifications like the Maginot Line, it is likely to be comparatively mobile. Consequently, there does not seem to be so much need for a second echelon to fulfil its old basic function. Therefore, consideration is now being given to the idea of using the second echelon as a force for exploiting the success of the original attack. In other words, the gap having been made because of the initial reverses inflicted on the enemy, the second echelon could be committed to a breakthrough operation, or something of that kind, with the job of pressing forward and exploiting the circumstances. The second echelon would thus be a body that a formation commander would have available to him in order to exploit his success.

On one final note, I would point out that on the strategic plane the Russians have two sorts of echeloning. In addition to the kind I have been describing, they also have 'strategic echelons'. To put things simply, one could regard all Soviet troops mobilized and deployed in the Group of Soviet Forces Germany (GSFG), Poland, Czechoslavakia, and Hungary as belonging to the first strategic echelon. The second strategic echelon starts with those troops not yet mobilized for battle and which lie back in Mother Russia and places further East. I only make this point because at a number of conferences that I have attended I have listened to a number of people talking about the need to clobber the second echelon. The point then arises: which second echelon are we talking about? And how does one know it, when one sees it? One thing we can be sure of, the commander of a Soviet second echelon is not going to carry a little flag saying, "I am the commander of the second echelon, please attack me". This makes it a difficult problem indeed.

Editors' Notes

1 Combat development is defined by the Canadian Forces as the process of determining the concepts, organization, materiel, doctrine, and training for use at some future date, and the orderly evolution toward the goals set by concepts. It is, in short, forward planning aimed at improving combat effectiveness.

2 Plumer commanded Second British Army, which successfully took Messines - Wytschaete Ridge to the south of Ypres in June, 1917. Called by Liddell Hart, "The only true seige warfare attack made throughout a seige war", it was the prelude to Passchendaele. See Charles Messenger, Trench Fighting 1914 - 18 (London: Ballantine, 1973) 99-10.

3 Professor Peter Vigor stated during questioning that the Soviets "tend to cling tenaciously to a well-tried and trusted product until they are absolutely ready to replace it with a new one". The Great Patriotic War practice of committing a second echelon only when the first echelon was more or less on its knees (which greatly facilitated the through passage of the former) is, in his opinion, thus likely to continue if such action is warranted. He argued that this doctrine had in the past served the Soviets well, and that they continue to believe that "people who try to dictate what is going to happen to a battle, once it has really got going, are likely to find themselves in a mess". In an associated vein, he also pointed out that the average Soviet army officer generally remains in appointment for a longer period than his Western counterpart, and that Soviet formation commanders in Europe are all hand-picked. The evolution of the Soviet officer corps over the last forty years has been such, moreover, that many officers' sons have followed in their fathers' footsteps and married other officers' daughters. In Professor Vigor's view, the similarity between today's Soviet forces in Europe and the 1914 British Expeditionary Force is thus striking,particularly insofar as those in posts of real responsibility know each other personally and may even be related;

they consequently retain a great deal of confidence in each other and "a sort of instinctive trust that, whatever the rules are, will allow the senior to give much more initiative to the junior than perhaps the regulations seem to permit".

CHAPTER THREE: MOBILITY ON THE BATTLEFIELD

by
Major C.R.M. Messenger
Sometime Fellow, Royal United
Services Institute for Defence Studies
London, England

On 15th September, 1916, thirty-two rhomboid shaped contraptions lurched their way across no-man's-land towards the German lines in the High Wood - Flers-Courcellette sector of the Somme. Many of the machines broke down before they could close with the enemy but, one, followed by British and New Zealand infantry, captured the village of Flers, one-and-a-half miles inside the German lines and made the best penetration of the day. A German officer captured that day described the use of tanks, as they were called, as "not war, but bloody butchery", while the Chief of Staff of the German Third Army in a more sober vein wrote: "The enemy ... have employed new engines of war, as cruel as effective. No doubt they will adopt on an extensive scale these monstrous engines, and it is urgent to take whatever methods are possible to counteract them."

The central issue of this symposium, the ability of small units to survive and fight on the modern battlefield is, in comparative terms, very similar to that which worried the German staff officer. Indeed, what happened on that now far off day spawned a new concept of land warfare, which, albeit refined through experience and advancing technology as the years have gone by, is with us still today and shows little sign of being replaced by anything radically different.

The intent of this presentation is to examine, in the broadest sense, mobility on the battlefield as it affects both the attacker and defender. We shall do this by first looking at the roles of the tank, which remains the cornerstone of mechanized warfare, then we shall cover other types of armoured fighting vehicles. This will be followed by an examination of the offensive versus the defensive and, finally, a look at some anti-armour tactical concepts. Liberal use will be made of historical examples, not merely because, being a military historian by trade, I am somewhat biased in this direction, but also because I firmly believe that past experience is a vital ingredient in the shaping of concepts of the future, and what more is history than past experience?

The battlefield of 1916, on which the tank made its debut, was one where the power of the defensive, because of the advent of the rapid firing weapon -- machine gun, magazine rifle, quick firing artillery piece -- had overwhelmed that of the offensive. The result of that was the deadlock of trench warfare and the realization that victory, should it come, was likely only to be pyrrhic. In its original concept, therefore, the tank was designed to physically break through the opposition's static defences to enable the infantry to bite sufficiently deep into the defence lines for the horsed cavalry of the time to fulfil their function of exploiting success by charging through the "G" in Gap. Thus, the tank's original role was merely one of infantry support. Then, on 24th April, 1918, outside the village of Villers- Bretonneux, came the first tank versus tank action, when three British Mark IV tanks, two of them "females" with machine guns only, took on three German A7Vs, resulting in one of the latter turning over in a sandpit. A second role for the tank, that of the destruction of other tanks, had arrived, but, because the Germans had very few tanks themselves at this time and they were used only occasionally, this role was not immediately recognized.

However, during the last two years of the war on the Western Front, one or two farsighted minds were thinking further ahead. Notable among these was Colonel J.F.C. Fuller, Chief of Staff of the British Tank Corps Headquarters, in France. On 24th May, 1918, Fuller wrote a paper, under the obscure title of The Tactics of Attack as Affected by the Speed and Circuit of the Medium D Tank. Colonel Fuller recognized that existing tanks did not have the speed or range to give them a role other than infantry support, but if such a tank was produced, the picture could be radically altered. He then went on to outline a concept of operations in which the tank would take a dominant part. This concept was later to be known as "Plan 1919" and propounded the thesis that the key to victory in an attack was the disorganization of the enemy's command and control system. Colonel Fuller envisaged the first phase of the attack being Medium D tanks, which would have a speed of 20 mph, two and a half times faster than any existing tank, combining with aircraft to attack enemy headquarters at army level and destroying supply and road centres. Then, with the enemy command thrown into confusion, the "Breaking Force" of heavy tanks, infantry and artillery would penetrate the enemy's front, being followed by the "Pursuing Force" of light tanks, lorry-borne infantry and cavalry, which would carry out the traditional exploitation role. This concept marked a radical leap forward in tactical thinking and represented what would be a foundation stone of modern mobile or mechanized warfare. Indeed, it is interesting to note that Phase One of "Plan 1919" bears a remarkable resemblance in its object and conduct to the latest Soviet concept of Operational Manoeuvre Groups.[1] Plus ça change .....

Thus, another and more wide reaching role for the tank, that of "shock action" had been recognized, and it was from this that mechanized warfare as we now know it would be evolved. The story of evolution of the theory of it in the 1920s and 1930s is too long to be described here. Suffice to say that military conservatism and financial stringency put a brake on the theories becoming fact in the Western democracies, while in Russia, after some spectacular progress had been made in the early Thirties, Stalin's purges temporarily withered the flower on the stem. Only in Nazi Germany was it really allowed to bloom into the blitzkrieg campaigns of 1939-41.

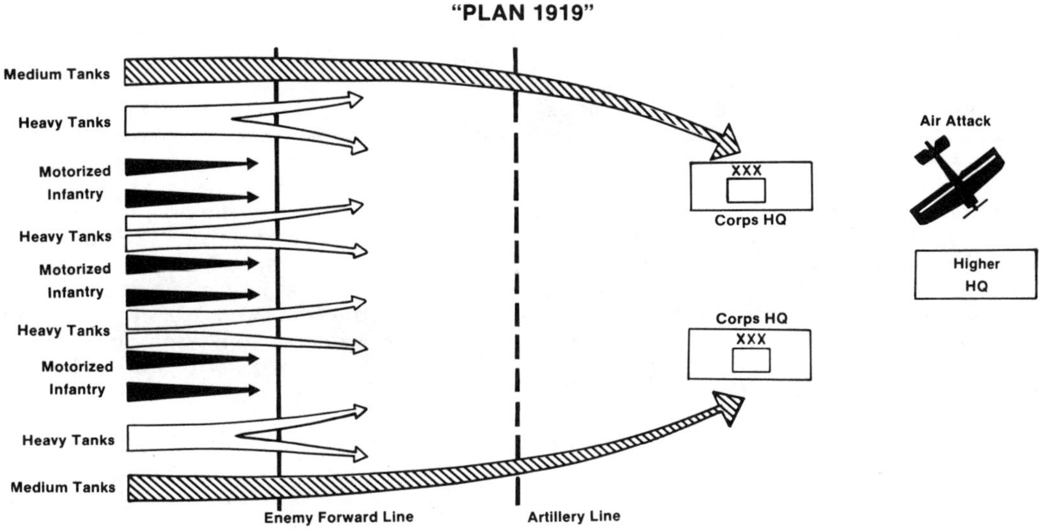

Strategical Paralysis as the Object of the Attack
Fig. 3-1

However, in view of the fact that the tank had now become a standard item in a nation's armoury, the anti-tank role began to assume ascendancy over the infantry support role, especially in the defence, and nowhere was this more apparent than in the fighting in the North African desert during 1940-43. Since then, by virtue of the fact that the tank has mobility, is armour-protected and possesses a very effective anti-armour gun, as well as being very flexible in having radio communications, the tank has been recognized as the principal tank killing weapon. By destroying the enemy's armour his mobility is severely restricted, particularly on the battlefield itself rather than behind it, and hence his combat power drastically reduced. This clearly illustrates why the tank continues to be a dominant weapon in the land battle. Nevertheless, the tank is not all-powerful and it does have limitations, the principal of which is that it cannot, on its own, hold ground. This is the main task of infantry, but their task of seizing and holding ground is made much easier through having the intimate support of tanks and this still continues to be a fundamental role of the latter.

The main reason why the German blitzkrieg eventually bogged down on the Russian steppes was because mechanized forces represented little more than the steel tip of the German Army. Thus, when "Barbarossa" was launched in June 1941, only 29 out of the 134 divisions involved were mechanized or motorized (by these two terms I mean that tracks and wheels were respectively the main means of mobility). The other 105 divisions were ordinary infantry, relying on their feet, with much of their logistic transport being horsedrawn. Reading General Heinz Guderian, that arch- practitioner of blitzkrieg, it is very noticeable how frustrated he became with those in command above him, both in the Low Countries in 1940 and in Russia, for their timidity when it came to giving him a free rein to exploit success with his panzers. The truth of the matter was that, spectacular though his successes were, with the enormous pockets of shocked and demoralized Soviet troops that he created, the mobility of his formations was so much in excess of those of the bulk of the German divisions that it put them way out in front and the dangers of being cut off and starved of logistic support became very real. Indeed, by the beginning of July 1941, two weeks after the campaign had opened, Guderian was, by his own admission, two weeks' march ahead of the main body. Indeed, how much easier life would be today if one knew that only the Warsaw Pact first echelon was mechanized, and the fact that this is not the case means that the well-worn principle of attack, maintenance of momentum, is very much easier for the Soviets to achieve than the Germans could.

Therefore, it is clear that an essential requirement for mechanized warfare is that the component elements must have the same degree of momentum, both on and off the battlefield. Thus, infantry and the supporting arms of artillery and engineers, as well as command elements, must travel in vehicles which, especially on the battlefield, can keep up with the tank and both vehicles must have approximately the same survivability. The Germans, within their panzer and panzer grenadier formations, did recognize this, and hence the ubiquitous Sonderkraftfahrzeug or Sdkfz 251 halftrack series, the workhorse of the panzer grenadier. Although open-topped, the Germans termed it Schutzenpanzerwagen or armoured personnel carrier, and indeed it was designed to be not just a battlefield taxi but to be fought from as well. The Allies, too, with the White halftrack and the Kangaroo, also developed tracked personnel carriers, which they used increasingly in the latter stages of the war.

In the Fifties the open top personnel carrier gave way to the true armoured personnel carrier (APC), which gave the infantryman overhead protection, but did and continues to have drawbacks. The infantryman in the back is little more than a passenger and apart from close range suppressive fire through weapons ports, in the sides and rear, cannot fight from the APC. Furthermore, with the limited vision the infantryman has in the APC he is likely to be disoriented on dismounting, and, if he has had to travel some distance in it, he may not actually feel too well either! Indeed, it is these latter points that have produced the controversy as to whether or not the infantryman should dismount in the final stages of an attack, but to go into this subject in detail is beyond the scope of this presentation. In defence, there was a further problem with the APC: armed only with a machine gun, it was of little use in the anti-armour battle, and hence, having deposited its passengers on their defensive position, it had to withdraw to the rear, there to wait like a radio cab until it was called to pick up its fare again. However, if the battle turned fluid there was and is a very real danger of the infantry not being picked up again.

Sdkfz 251: First Successful APC
Fig. 3-2

Thus, thought turned to making the APC into something more of an armoured fighting vehicle (AFV), and the result was the mechanized infantry combat vehicle or MICV. The first of this type to appear on the scene was the Russian BMP[2] in 1967 with its now very familiar 73mm smoothbore gun and Sagger anti-tank guided weapon (ATGW) launcher. This was followed shortly after by the West German Marder, more lightly armed with just a 20mm cannon and two machine guns. Recently, with the American M2 Bradley, the French AMX-10P and the British MCV-80, a whole host more MICV's are appearing. The main feature of the MICV is that it can take its place in the firing line and, thereby, give the infantryman more combat power. Those MICVs like the BMP and the Bradley, which are armed with ATGW, can take on main battle tanks but, essentially, the MICV is there to deal with similar or lighter armoured fighting vehicles, thus enabling the tanks to concentrate on their opposite numbers. It also means that, when its infantry have dismounted, the MICV can stay close to them, giving them more chance of being picked up and hence greater flexibility. There is, however, a cost penalty. A MICV is many times more expensive than a straightforward APC and many countries are doubtful that they can afford all that they require. Also, many infantrymen believe that its complexity might result in a degredation in basic infantry skills. This means, certainly in many NATO armies, that the APC will soldier on for some years to come.

Bradley Infantry Fighting Vehicle
Fig. 3-3

Artillerymen often claim that it is their arm that is the "queen of the battlefield" and in many ways this cannot be disputed. In conversation with an officer who fought in the Falklands he explained that one of the main, if not the major, lessons he had learned was never to move without artillery support. Indeed, one of the reasons why attacks across trenches during the Great War often petered out was due to an inability to move artillery across shell-torn ground in time to support the next phase of the attack, or, to block the inevitable counter-attack. Similarly, in 1918, Marshal Philippe Pétain, so often dubbed the arch-proponent of static defence because of his enthusiasm for the Maginot Line in later years, saw a means of countering the new German attack tactics, which had given the British such a fright in March and April of that year. Let the sturmtruppen come on at such a rate that they outstripped their artillery, and then they would be easy pickings. Indeed, it was a classic form of mobile defence but the very idea was anathema to most French generals, imbued as they were, especially after Verdun in 1916, with the notion that to give up one square metre of la belle patrie voluntarily, without a fight, was tantamount to treachery. Nevertheless, that summer of 1918, when the Germans turned on the French, Pétain's concept did slowly take hold, and by July the German Army was a spent force offensively.

The lessons from this were obvious and, indeed, were old lessons being re-learned. Artillery must have the mobility to keep up with infantry and armour. Yet, curiously, during the between-the-wars years little attempt, apart from towed motorization, was made to give the guns the necessary mobility. True, there were a few experiments in producing self-propelled (SP) guns (the British had even produced some called "gun carriers", mounting a 60 pounder or 6-inch howitzer on a tracked chassis in 1917), but even the Germans did not appreciate the need to integrate self-propelled artillery with their mechanized formations until 1940, and then it was anti-tank rather than artillery per se. Indeed, it was the Americans who led the way with their 105mm gun mounted on the M3 tank chassis, which first saw action at El Alamein in October 1942 (the Germans did not produce any until the following year). Nevertheless, by 1945 the self- propelled artillery piece had become a necessary element in mechanized warfare, except in Russian eyes. Here, in spite of traditional belief in the dominance of artillery on the battlefield, the first SP models, M-1973 and M-1974, did not make an appearance until the early Seventies.

Soviet 122mm SP Gun
Fig. 3-4

Having now examined the traditional triad of mechanized warfare -- armour, mechanized infantry and SP artillery -- I want to go on and weigh the scales between the power of the offensive and that of the defensive. As I have already said, the advent of the tank meant that the pendulum swung back in favour of the offensive. Yet, apart from the "tank enthusiasts", between the wars there was only very slow realization of this. Indeed, it was the horrors of trench warfare that had imprinted themselves on the minds of most, rather than the methods used to break it. Thus, the conservative view still held that defence could be overwhelming if it was built around strong enough static fortifications -- hence the Maginot Line, and later the Czech defences and even the German Siegfried Line. Somewhat erroneous reports also came out of the Spanish Civil War. As one French military writer wrote: "The antidote to the tank has been found. It can no longer count on technical surprise. It fights on equal terms with anti-tank defence, and the theories advanced on the devastating power of the 'Panzer divisions' and other armoured formations, employed in isolation, have been overtaken by events".[3] The weapons were certainly there to counteract the tank, but what the writer missed was the need for their users to be able to think and react in "tank time", as the events of 1939-41 showed only too clearly, as well as confirming the power of the offensive over the defensive.

Yet, every weapon eventually has its antidote. In the case of the tank, a gun-armour race developed in the same way as the Dreadnought race of the 1900s. However, one curious result of this was seen in North-West Europe in 1944-5, where the Germans showed that they were winning the race with their Tigers and Panthers over the Allied Shermans and Cromwells. The result of this was that the tank itself became stronger in defence rather than offence. Just to illustrate what I mean, here is part of a troop leader's account of Operation "Goodwood",[4] in July, 1944:

> "Bill, my squadron leader, gave his orders over the wireless in an excited voice, but they were perfectly clear. They were to be his last... As soon as Bill had completed his orders, we started to move slowly forwards. Within a few seconds Peter Robson's tank was hit by an .88 and went up in smoke. All the crew baled out safely. The rest of the squadron moved on and I could hear Jock Addison reporting a Panther on the outskirts of the village, which he was trying to engage. This was almost the last coherent message to be heard from the rest of the squadron. They were all brought under heavy and accurate fire and within a matter of minutes about five tanks were on fire and another three out of action...Everything was now in a state of chaos and the squadron on the right was also being rapidly engaged by the enemy and losing tanks rapidly. A general withdrawal in the direction of the railway line took place, but it seemed only a matter of time before the rest of the regiment was written off."[5]

There was, too, the question of terrain. The high banks and hedgerows of the Normandy bocage, and indeed the closeness of much of the countryside of North-West Europe as a whole made the tank particularly vulnerable, not just to other tanks, but also to the infantryman on his feet with a hand-held anti-tank weapon such as the Panzerfaust, PIAT or bazooka. This was particularly difficult for the Desert veterans to adjust to, but did make them appreciate the value of their accompanying infantry. This, and the fact that Allied numerical tank strength was far in excess of their enemy, enabled them to get forward.

The Russians, too, had become firm believers in the old Napoleonic dictum that "God is on the side of the big battalions", and began again, much as they had in the early Thirties, to mass large mechanized forces. With the opening of the Cold War in the late 1940s and the formation of the two opposing alliances, NATO and the Warsaw Pact, the concern of the Western democracies turned quickly to the problem of the defence of the continent of Europe.

In the event of war in Europe, the Russians, cloaking the concept under the term "the counter-offensive" that implied for propaganda purposes NATO striking first, envisaged punching through the NATO defences with large armoured forces, with the aim of carrying on until they reached if not the Rhine the English Channel. As for any attack, they relied and indeed set great store in maintaining momentum to carry them through to their objective in a matter of days. The NATO concept, when faced with this, became quickly reliant on the tactical nuclear weapon. Time could be bought by fighting a delaying action from the inner German border back to the main defensive position, which was based on a natural obstacle comparatively well back. Those of you who were in Germany with the Canadian brigade at the time will recall that for I(British) Corps this was the River Weser. Here more time would be bought while the enemy tried to cross, but it was accepted that he eventually would succeed. Then it was a question of canalising him into an area designated a "nuclear killing zone", where he could present a large enough target for a tactical nuclear attack to be made.

Conscious that NATO would never be able, for political and economic reasons, to come anywhere near matching the Warsaw Pact man for man and tank for tank, there was a certain feeling of fatalism when it came to considering the defence of Europe in strictly conventional terms. While NATO knew that it could build better tanks, and they were still the prime anti-armour weapon, their numerical inferiority still meant that they were likely to be engulfed before they had destroyed enough T54/55s, and later T62s, to halt the enemy. Hence, the only way to push the pendulum back in favour of defence was the tactical nuclear weapon. However, one anti-armour weapons development in the late 1950s did begin to cause some rethinking.

The 105mm tank gun as on Centurion and M60 had an effective tank killing range of 2000 metres, but the advent of the ATGW in the form of the American Shillelagh and British Malkara, among others, extended this range to 3,000 metres with a high probability of a kill. Although terrain studies of West Germany had shown ranges of visibility to be in excess of 2000 metres only ten per cent of the time, careful positioning of ATGW could take advantage of its maximum range, and meant that the Soviet armour could be taken on farther out, which in turn meant more time to fight the battle of attrition with more chance of destroying more enemy armour. This altered the position of the pendulum in the conventional war sense, but the infantry could only play little part in this battle. Yes, they could hold ground and engage enemy armour with their recoilless anti-tank guns and hand-held rocket launchers at the closer ranges, especially in thicker country, but there was a dangerous tendency to regard them, particularly by those of a different cap badge, rather like a sandcastle as the tide comes in, with the armour of both sides being the tide, which ate away at the castle with impunity.

The tank also improved its anti-armour capability at this time, especially in terms of improved fire control systems. The use of rangefinders, stereoscopic or ranging machine guns, meant more accurate opening ranges of engagement, with a better chance of a first or second round hit, and this improved its capability as an anti-armour weapon still further, thus contributing to the increasing power of the defence. Some were even tempted to think that infantry were almost an embarrassment on the European battlefield, even though they might be mechanized, and that a future war would be won by the tank, with, it was grudgingly admitted, some artillery support. However, the advent of new weapons in the 1960s seemed at the time to threaten the future of the tank itself.

The first was the helicopter, which had been initially blooded in the Korean War. It really began to come into its own though in Algeria, and later, to very much larger extent, in Vietnam where the helicopter gunship was developed. With ATGW now very much established, it was a logical step to fit this to the attack helicopter and trials, both in the United States and in Europe, seemed to indicate that it was a decisive tank killer. The climax came in early 1972 with some spectacular American successes against North Vietnamese armour during the final invasion of the South. There was by this stage a very strong helicopter lobby, especially in the U.S. Army, and some of the more extreme of its members even propagated the theory that the helicopter was about to take over from the tank. But, if this made the "all tank" school pause, events the next year really made them think.

The one country which, by the early Seventies, did have wide practical experience of armoured warfare, albeit in desert rather than European terrain, was Israel. After her victories in 1948, 1956, and 1967, she could afford to be a little complacent. Nurtured on Liddell Hart's strategy of the indirect approach, which experience had shown them was best achieved on the ground by fast moving armoured columns, and the fact that their Centurions and Pattons, with superior crews, outmatched the Arabs' Warsaw Pact armour, they became convinced that the tank on its own was a battle winner. Egypt continued to be seen as the main enemy, and in order to defend their gains in Sinai, the Israelis constructed defences on the east bank of the Suez Canal, which were known as the Bar Lev Line, and planned that these should delay the Egyptians long enough for them to concentrate their armour to send them reeling back. The Egyptians, of course, in October 1973 managed to catch the Israelis by surprise and breached the Bar Lev Line. While the armour was being concentrated, the Israelis launched unsupported tank attacks against the Egyptians, and one who took part in one of these recounted:

> "We were advancing and in the distance I saw specks dotted on the sand dunes. I couldn't make out what they were. As we got closer, I thought they looked like tree stumps. They were motionless and scattered across the terrain ahead of us. I got on the intercom and asked the tanks ahead what they made of it. One of the tank commanders radioed back: 'My God, they're not tree stumps. They're men!' For a moment, I couldn't understand. What were men doing standing out there -- quite still -- when we were advancing in our tanks towards them? Suddenly all hell broke loose. A barrage of missiles was being fired at us, many of our tanks were hit. We had never come up against anything like this before...!"[6]

This culminated in the complete destruction of the Israeli 190th Brigade before they realized what they were up against, In truth, the Egyptians were extensively armed with the Russian Sagger ATGW, which, of course, can be operated with ease in the ground role, and the only way in which the Israelis were able to counter this was by incorporating infantry with their armour. Indeed, so firmly had the lesson gone home that their new homegrown main battle tank, Merkava, combines the role of armoured personnel carrier with that of a tank.

Ground Mounted Sagger
Fig. 3-5

The infantryman's ATGW and that of the helicopter marked a growing proliferation of this form of anti-armour weapon on the battlefield, and the pendulum began to move further towards the defence. However, in spite of this increasing dominance, which encouraged "the tank is dead" school, calmer minds prevailed, and during the early 1970s grew the concept of "the family of anti-armour weapons". Here, whether the weapon was long, medium or short range, aerially (and this includes fighter ground attack as well as attack helicopters) or ground delivered by guided missile launcher, tank gun or recoilless tube, man-portable, track or wheel-borne, each had its role to play in the anti-armour battle. The strengths and limitations of each were recognized, and the part that each was to play made the most of the former. Indeed, commanders at every level became rather like orchestra conductors.

One significant result of all this is a changing role for the tank on the battlefield, at least in NATO's eyes. Previous to the flood of

ATGW, the tank, of course, as I have said, was the main anti-armour weapon. However, this meant putting most of the tank assets in the "shop window" so that they could fulfil this task. This meant that commanders had limited numbers only available for counter-stroke purposes. Now, more armour can be held back and more can be made of what NATO commanders see increasingly as vital to the concept of conventional operations for the defence of Western Europe.

Two other recent technological developments also act on behalf of the defensive. The first is the remotely delivered minefield, which is a tremendous asset to a commander, firstly when faced with a threat from an unexpected quarter, and also in that he has greater flexibility by virtue of the fact that he does not have to rely so much on pre-laid minefields which always have the danger of restricting his own options. Secondly, there is the advent of the precision guided munition (PGM). Spearheaded by the American Copperhead, this means that artillery can engage individual tanks indirectly, with merely an observer forward with his laser designator. An interesting development of this is the precision guided mortar (the French firm of Thomson-Brandt are developing a 120mm one and there is the American GAMP) which, with the high attack angle of the projectile, means that an area in which tank designers have in the past been able to save armour, the top, is now under threat. Thus, there are two more instruments to add to the orchestra, and we can place another two weights on the defence side of the scales.

Although the multiple anti-armour threats appear to provide for a gloomy future for the tank, developments in the latter have not stood still. More sophisticated fire control systems using laser range finders mean faster target acquisition and still greater chance of a first round hit. The latter has also been improved through the introduction of fin-stabilized tank gun ammunition, and the ability to use hotter charges, which results in higher muzzle velocities and more accuracy. Also there have been very significant strides forward in the field of protection. A new form of composite armour has been developed that is particularly effective against the hollow charge warhead which, of course, is used in ATGW. It is claimed that the new composite armour is three times as effective as steel armour of the same weight. Such armour is likely to be impervious to certainly the smaller missiles, which, of course, puts a question mark on the infantryman's tank killing capability in the future, and means that "the terror of the tank" is still with us to an extent. Nevertheless, there are enough anti-armour "goodies" around to put up an effective defence, but what sort of defence, and how does the small unit fit into it?

There are, at the moment, four schools of thought in NATO as to how the defence of the Central Front in Europe should be conducted in the future, and all of them use the assumption that the policy of forward defence, as evolved for political reasons in the late Sixties, will remain with us for the foreseeable future. I have dubbed the first the "Maginot" concept in that it argues that fixed fortifications should be constructed. These could act more as a deterrent than anything else and, although it might appear far-fetched, there are degrees of fixed fortifications. Consider, for instance, the Soviet view of the encounter battle, that of catching NATO formations before they can fully deploy from their barracks -- prepared positions

Leopard 2 Main Battle Tank
Fig. 3-6

become attractive as they would prevent this by slowing down the rate of the enemy's advance. Such fortifications would, of course, take time to construct even given the political will. Then there is the partially static defence which urges fixed positions and obstacles to wear the enemy down with a mobile reserve being then launched to drive him back across the border. The third type is really a variation of the second with less being used statically and more kept in reserve. Finally, there is the school which argues that defence of ground per se in what is bound to be a highly mobile battle is out of context and that a pure mobile defence should be conducted (rather in the same way as a battle at sea) with highly mobile forces manoeuvring independently on the ground. My own view is that Options 1 and 4 are somewhat extreme, and that the static/mobile defence as put forward in Options 2 and 3 is likely to be the way in which we will fight the battle.

One point that is worth mentioning here, and which should be borne in mind, is the growing conurbanization of West Germany. One only has to compare a 1:50,000 map of a particular area today compared with that of twenty years ago to note how not just the major towns are becoming more and more sprawling, but the villages as well. In fact, compared to Canada, and even to Britain, the density of villages is remarkably high. This in itself favours the defence in that it restricts the Warsaw Pact more and more in "beating about the North German Plain", if I can borrow an expression from Field Marshal

Montgomery, although, of course, he was looking in the other direction. This acts very much in favour of Options 2 and 3, and we have, as a result, in recent years come to think much more seriously about fighting in built-up areas than we did previously. Nevertheless, there is still much very defensible terrain away from these areas, and we should not allow ourselves to become myopic.

The likely layout of the defence will be as a lattice work of strongpoints and obstacles, both artificial and natural and has been dubbed framework defence. It could, in the case of Option 2, be designed to prevent the enemy getting all the way through it with the mobile reserve not necessarily having to be used. In other words, the decisive battle would be fought within the lattice work. For Option 3, on the other hand, it is the counterstroke by the mobile reserve that will be decisive, and hence the lattice will be designed to guide the enemy into an area where he can best be counter-attacked. Nevertheless, it will be so laid out that the enemy will be relatively weak and groggy by the time he comes to face the riposte. In either case, I do not foresee the defence within the lattice as being entirely static. For a start, provided that time allows, alternate strongpoints and obstacles will be prepared but not necessarily occupied. This would cater for the enemy coming in an unexpected direction, and it would be a question of quickly moving to occupy them. There will also be the case where a crucial strongpoint, say on a very dominant piece of ground, falls. This might well have a dangerous "knock on" effect if allowed to remain in enemy hands, and hence there will be a need to recapture it quickly. This will require the defenders of the lattice to have the necessary mobility with them to carry out either of these two tasks.

In terms of strongpoints, these would come in two types, urban and rural. In respect of the former, it should be remembered that every building in West Germany must, by law, have a cellar and windows fitted into it so that they can be used as weapons slits. Hence, there is already the basis of a strongpoint there. They will, of course, be very much the province of the infantryman on his feet, and will provide plenty of scope for the hand-held anti-armour weapon. Within the main towns, they are likely to be the province of the local German territorial forces, although it should be borne in mind that the enemy first and second echelons will not want to get bogged down in the cities, thereby losing momentum. The suburbs should nonetheless provide a good base from which to launch nuisance attacks on enemy armoured columns as they try to bypass.

The rural strongpoint will be located where it has good fields of fire and dominates an approach or approaches, and mechanized forces, both armour and infantry, will be employed as opposed to non-mechanized troops as found in the urban strongpoints. Here the enemy, desperate as he is to press on where the going is good for his armour, will be forced, if the positions are well placed, to at least drop troops off to mask them, if not actually attack them. A classic example of this is the German Ardennes offensive of December, 1944, and the influence of Bastogne and other strongpoints on the outcome. Thus, the enemy will be slowed down and will have his strength dissipated. The strongpoints themselves will be dug and occupied by infantry, and here it is worth pausing for a moment to consider the business of digging on the mechanized battlefield.

Mechanized infantry sitting in the back of a MICV or APC may have combat potential but that is a wasted asset if they never get out of the back. Indeed, where it comes to the point of holding ground, if it is to be done effectively, an element must be dismounted so as to properly cover the approaches, particularly those which lie in closer country, and use the weapons that it has been given to fight dismounted. Apart from the nuclear aspect, which of course gives motivation to dig and which I purposely have not touched on in this presentation, artillery will be very active, and certainly from what we know, the Soviets still place great reliance on it. Hence, unless dismounted infantry dig, they are going to be very vulnerable and unable to do their job. In fact, it may well be a question, if there is time and if the resources are available, of digging positions for tanks and MICVs as well as ATGW launcher vehicles in order to give them good well-protected fire positions. Even the mechanized infantry in the mobile reserve will, on recapturing ground, have to dig-in to effectively consolidate their hold on it. There will, more often than enough, not be the time to dig elaborate trench systems, but, as he has always done, the infantryman will dig enough to give himself a feeling of some security.

As for the battle itself, I would see fighter ground attack (FGA) taking on the follow-up echelon, while precision guided artillery and long range ATGW begin sniping at extreme visible range. Once inside 3,000 metres, or effective tank range, the tanks will join in, and then medium range ATGW. MICVs will concentrate on their opposite numbers, while dismounted infantry will concentrate on AFVs that get within close range, as well as dealing with dismounted enemy infantry. In the event of a sudden threat developing, the attack helicopters, with their quick reaction time, will be called in. Remotely delivered mines could also help at this point. Supporting conventional artllery will also play its part. The main idea is that every arm has to participate in the battle; the anti-armour battle is the business of all arms.

In order to draw the threads together of this traverse through the mechanized battlefield of the last 65 years or so, as well as that of the future, I would like to stress several points. First, I do not believe that mechanized warfare is going to go away in a hurry. There are an increasing number of informed minds who believe that the helicopter has a greater role to play in it, and advocate the use of air-mechanized forces -- Brigadier Dick Simpkin is one, and General Dr. von Senger und Etterlin, the present Commander-in-Chief, Allied Forces Central Europe, is another. However, they agree that although this will add a new dimension to the European battlefield and assist in maintaining mobility, it will not be at the expense of the ground dimension. The combined arms concept will remain with us but, with military technology ever advancing, the battlefield is becoming a more and more complicated place. This means that commanders at all levels, from section upwards, are going to need much greater breadth of thinking, a deeper understanding of the strengths and limitations of the weapons systems under their command, and an increasing ability to think and react even more quickly. One of the major keys to the German successes of 1939-41 was the ability of their armoured commanders to act more quickly than their opponents, and the Warsaw Pact will hope to do the same, for it is this that brings about dislocation of command. Thus, we must ensure that our speed of reaction outmatches that of our potential enemies. Finally, a weapons system, however sophisticated, is only as good as its operator, and it is still, to put it rather simplistically, the ability of the infantryman with rifle and bayonet to get up out of the trench and go forward when the fire is hot that will bring about ultimate victory. Only thorough training giving the ordinary soldier confidence in himself and his weapons and the right motivation will enable him to do this.

Editors' Notes

1 See C.N. Donnelly "The Soviet Operational Manoeuvre Group: A New Challenge for NATO", International Defense Review, No 9 (1982), 1177-1186; and C.J. Dick, "Soviet Operational Manoeuvre Groups: A Closer Look", International Defense Review, No 6 (1983), 769-776. "Plan 1919" was also previously called "Strategical Paralysis as the Object of the Attack".

2 For Boevaya Mashina Pekhoty, which means infantry fighting vehicle.

3 Commandant Cailloux, described as one of the ablest French technicians, writing in Revue d'Infanterie (Paris, 1938). Though tanks were no longer a technical surprise, new methods of using them permitted the attainment of tactical surprise. Miksche, Blitzkrieg, 101.

4 Intended by Field Marshal Montgomery to "write down" German armour, "Goodwood" launched General Richard O'Conner's XIII Corps, of three armoured divisions, in a major attack east and south of Caen on a two kilometre front. Although supported by over 2000 aircraft, the operation actually "wrote down" more British armour than German; in 72 hours XIII Corps incurred more than 300 tank casualties.

5 Charles Messenger, The Art of Blitzkrieg (London: Ian Allan, 1976), 200.

6 Insight Team of the Sunday Times, Insight on the Middle East War (London: Deutsch, 1974), 106-107.

CHAPTER FOUR: THE MEANING OF MOBILITY

by

Brigadier (Retd) R. E. Simpkin, OBE, MC
ELS Consultant Linguists Limited
Over Deanshaugh, Scotland

The following presentation was delivered in Ottawa to a group of officers from the Cameron Highlanders of Ottawa and National Defence Headquarters. It is particularly pertinent to the "manoeuvre versus attrition warfare" doctrinal debate currently on-going in U. S. military circles.

From the beginning of history, rival philosophies, schools of criticism, cosmological theories -- even in many cases religions -- have co-existed with each being accepted as valid within its own frame of reference. But war is a very physical business by definition and by nature. And land warfare is tied to a unique physical reality produced by the combination of topography and technology. So it is not easy to accept that two conflicting doctrines on the conduct of war can both be valid.

When I was a serving officer striving for consensus between Allies over tank policy, I inclined to Field Marshal Lord Carver's view of a gradation between the "apostles of mobility" at one extreme and what one might call the "prophets of position" at the other.[1] But the absence of responsibility and the cathartic effect of writing for publication have left me in no doubt that there are in reality two clear-cut and diametrically opposed schools of thought.

The Anglo-American school is based on a literal interpretation of von Clausewitz' "concept of destruction", and on firepower as the principal instrument of imposing the commander's will. The other, pioneered in modern times by Fuller and his colleagues, given shape by Liddell Hart, voiced by Charles de Gaulle, and implemented by two generations of Germans and of Russians, allows "disruption" as an alternative to destruction and is based on <u>the "movement of masses" and on the dynamic forces thus created</u>. In this regard, Major General A.A. Strokov, writing in the late seventies in a book called <u>Military History</u>, had this to say:

> "The character and methods of manoeuvre in the course of attack have tangibly changed. While earlier the basic goal of manoeuvre was striving quickly and in an organized manner to

transfer forces and place them in a more favourable position in relation to the enemy, now to this has been added the necessity for rapid advance into the depth for the swift use of the results of the nuclear strikes and for seizing enemy objectives."[2]

As this remains the crux of the whole matter, in my opinion, I would like to put down three markers before expanding upon the view I have formed to this point. First, it may be helpful to clear our minds on the conventional understanding of the term "mobility" as applied to land warfare.

Strategic mobility, the ability to get into military business, is something that the heyday of the NATO centre conveniently allowed us to forget. But events over the past few years, in the Persian Gulf, the Falklands, and other foci of destabilisation, have reminded us of it very sharply. Strategic mobility is basically, I suppose, about transportability and logistic supportability. But we have to remember that it also means the basic suitability of vehicles to the terrain in which they are to operate.

Operational mobility, and here, of course, I am using the word "operational" in its German or Russian sense, is, figuratively as well as literally, a foreign word to us. It means the ability to move fast over considerable distances -- say of one hundred to three hundred miles -- on roads and good tracks. And, of course, to arrive fit to fight.

If tactical mobility still means anything at all, it means the ability to cross water obstacles quickly and reliably, or in other words, to swim. Those who have read the description of the assault crossing in the 1967 Exercise "Dnieper" in the defector "Viktor Suvorov's" first book,[3] will know just what I mean. This sets an upper limit of about forty tons of vehicle weight.

Now we come to the joker in the pack, or perhaps the ace in this conventional suite, namely, trafficability. This again is a matter of all-up weight. Even within the Federal Republic of Germany (FRG), whose modern road system was largely designed to NATO requirements, each assigned division has only one east-west route of class sixty or above. Class forty would make movement a good deal easier. But to really open up even this most developed of terrains, we have to get down to class thirty, twenty-five tonnes plus -- the original target weight of Leopard 1 and a figure engraved on many a German heart.

I have come to see battlefield mobility as a very dubious concept. It not only conflicts with the aspects of mobility on which we have already touched , but it is meaningless in itself, and is composed of two quite different and to a large extent mutually contradictory elements: mobiquity, the ability to cross bad going; and agility, the reduction to the minimum of the time a vehicle is exposed when moving or firing. Agility has demonstrably become the most important single factor in the survival of armoured vehicles on future battlefields.

Mobiquity, on the other hand, conflicts with just about every other mobility factor, not least trafficability. Crossing swamps needs a wide vehicle; weaving through forests a narrow one. This is a trade-off that the user is going to have to look at very carefully. It is extremely interesting that the Soviets should have allowed the nominal ground pressures of their key armoured vehicles to go right up.

My second marker is historical. If you look at the pictures in cavalry messes the world over, and study the dominance of the horses' shoulders, you will be in no doubt that the fighting power of a cavalry charge lay in movement. The opposing infantrymen or gunners were knocked over and crushed. The lance or sabre -- what we now call firepower -- were secondary. The same is true of the era of fast chariots. The blades on scythe chariots were awe-inspiring but they were secondary to the impetus of the charge.

My third marker will serve to introduce the mechanistic approach that I am going to use. When an engineer designs a vehicle, say a simple one like a half-ton cargo trailer, he first considers it as standing still and works out how thick each piece needs to be for the thing to support itself and stay in one piece. The dynamic loads, to which it is subject when it is moving, are far harder to calculate. In a simple case like this, he might double his static figures; and this is before he applies whatever safety factor he is using. Dynamic forces are thus a physical reality. The question we hate to answer, of course, is whether these dynamic forces still truly contribute to fighting power in an age when contact between opposing armies takes the form not of a clash of bodies or hulls, but of an exchange of fire.

The confusing thing about this doctrine of disruption and the movement of masses is, I believe, that there are four entirely distinct physical elements to it, which I should like to consider before moving on, as I shall later, to consider the psychological aspect.

We first need to think of a substantial armoured force advancing across country at high speed and density. Its forward elements may be -- though if they are Soviet they probably will not be -- fully deployed and practising fire and movement. But the rest is definitely in column or semi- deployed and rolling; it adds up to a considerable moving body. In physical terms the body possesses potential or stored energy in the shape of firepower; kinetic energy, meaning simply the energy of movement, which is increased fourfold by a doubling of speed; and momentum, or resistance to change of speed or direction, which is proportional to both speed and mass. In a word, this body is going to take a lot of stopping.

There are, in addition, four important points to note about this force. First, to the extent that these dynamic forces translate into fighting power at all, they do so purely at tactical level. Second, inherent momentum makes this force's behaviour predictable; changing its speed or direction is very nearly as difficult for its own commander as it is for the defender (it is distinctly short on the randomness that the Soviets rate so high). Third, time and space alone dictate that the advancing force can only exploit its potential energy by sacrificing some of its dynamic power; in other words, by slowing down or stopping. Fourth, to stop the advancing force the defender needs, first, an intense concentration of firepower in time and space -- above all in time -- to nullify the dynamic forces; then a short, sharp counter-stroke to throw the enemy into reverse; in practical terms, to make the follow-up forces concertina.

This, I suggest, is where the "hammer and anvil" defence tactic comes in. When David went forth against Goliath, he used a stone from his sling -- translational movement. David might equally well, given the right ground, have tied his stone to the end of a rope and whirled it round and round -- rotational movement. Fig. 4-1 shows how, if the anvil holds, it gives the defender a secure space in which to swing his hammer and thus generate these dynamic forces.

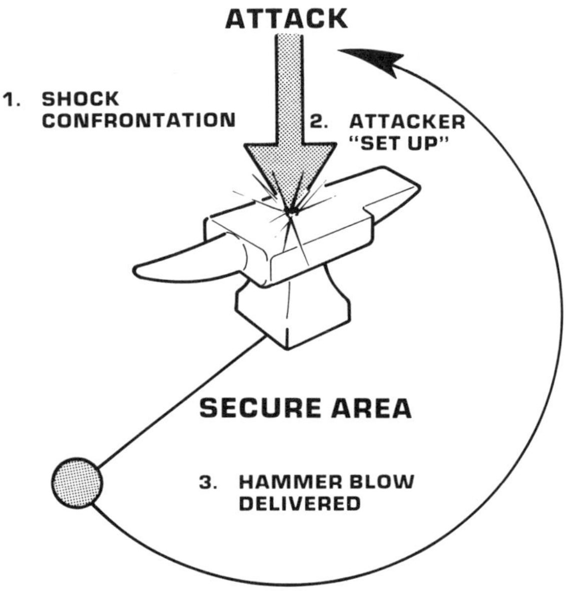

Hammer and Anvil Tactic
Fig. 4-1

The other three physical elements of the doctrine of movement of masses are, I suggest, far more important because they exert influence at operational or even strategic level. At this point I would like to emphasize an obvious but often overlooked advantage of the attacker and to highlight the associated Soviet concept of operational depth. This concept is one which continues to dominate the thinking of leading Soviet authorities such as Marshals O.A. Losik and A. Kh. Babadzhanyan on armoured warfare,[4] despite the current emphasis on the "all-arms battle".

You can see from Fig. 4-2 that the attacker who succeeds in penetrating and breaking through the defence progressively gains, initially at tactical level and then at operational, interior lines. In other words he gains, as well as the initiative which the attacker by definition possesses, a lateral freedom of action -- the ability to switch his line of advance over an increasingly wider arc. This is the logistic concept of interior lines making a rather dramatic appearance in combat dress.

The Soviet concept of operational depth means the depth from which a move capable of having effect at operational level can be launched. In the Second World War or even the Seventies concept of mobility, we might regard this depth as thirty-five to fifty kilometres. For this decade and the next, we might double this figure. Evidently the concept is geared to two factors. One is the depth which a defender, weak enough militarily to be on the defensive, might reasonably be able to hold in strength. Any advance beyond this will force the defender to react at operational level, that is, at corps level or higher. The second factor is probably the depth at which, looked at geometrically as the Soviets do, the arc of lines of advance widens enough to cross the defender's higher tactical -- say divisional -- boundaries, thus again calling for a reaction at higher level. This is also the depth at which the "tank force" proper, the third wave in what we might now call the classical threat scenario, passes through and cuts loose.

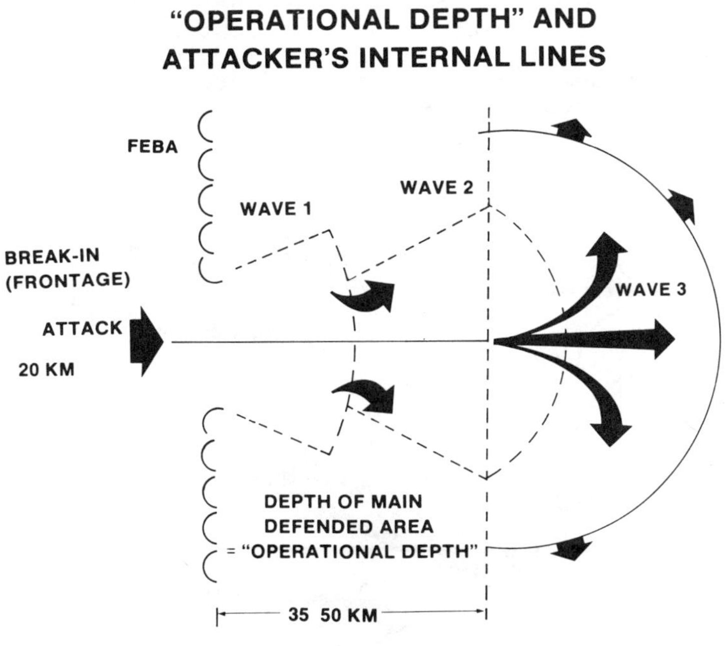

Operational Depth
Fig. 4-2

The second physical element is quite simply movement, the principle of the lever arm, which is illustrated in Fig. 4-3. Behind the Inner German Frontier (IGF) and the Rhine and lower Meuse rivers are depicted the initial Warsaw Pact strategic objectives on the Bonn-Rotterdam axis. The solid goose-eggs represent the main defending forces so that their centre of mass lies somewhere along the broken line. On the enemy side, the small blob represents a mobile force -- the "mobile column" of the Red Army in the Second World War, now renamed an operational manoeuvre group (OMG) -- which has covered a good deal of the distance between the defender's centre of mass and the pieces of real estate he is out to defend. In other words, this blob is a small mass with a very long lever arm. Because of this, it exerts a moment, or disruptive effect, quite out of proportion to its size. The blob can be enlarged from the air, but I incline to the Soviet armoured view that, in conventional warfare, the lever arm must actually have been traced out -- a mobile force must have broken out and advanced. I admit, of course, that much of this sounds like a mixture between a primary school class and a visit to the optician. But I suspect it is failure to see the thing in these basic terms that has landed us all in the slough of confusion in which we undoubtedly wallow.

And so to the third aspect. This blob can exert still greater pressure in one or both of two ways, and this is precisely the classical concept at which Major General Strokov was getting. It can be reinforced along its own path or by air; or for that matter by sea, a thought very much in the front of Soviet minds these days. Or it can lengthen its lever arm by pressing on. And so we come back to a truly dynamic quality: the rate of change of moment, in other words tempo. The more I peer into these depths, the more convinced I am that this is the key characteristic -- the one that decides whether or not the defender, however hard pressed at tactical level, retains or loses control of the operational situation.

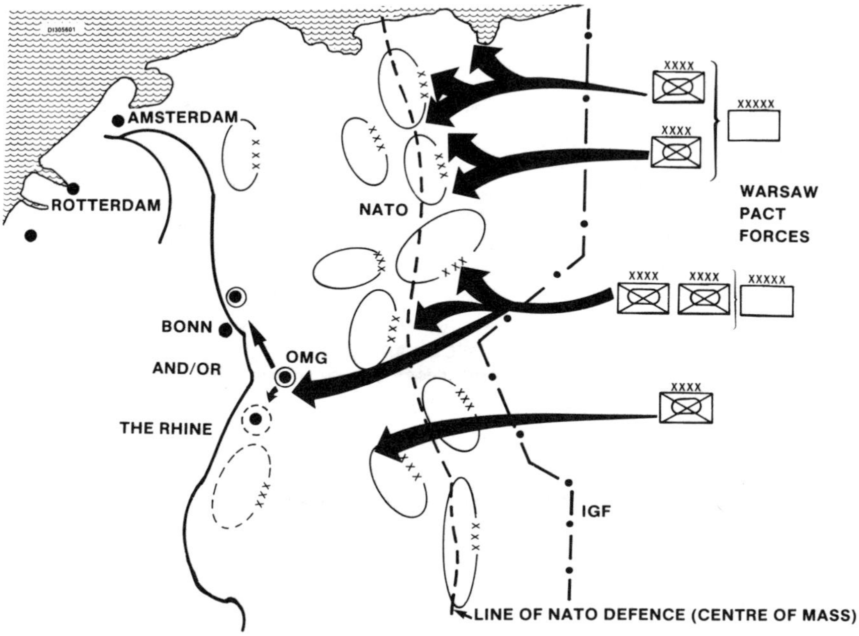

OMG: Principle of the Lever Arm
Fig. 4-3

For reasons I perhaps need not fully explore, the actual outcome of modern air and sea contests is likely to lie within a fairly narrow band -- a "short bracket" -- astride the most probable result statistically predicted from the actual physical situation. By contrast, because ground is both man's own element and an infinitely complex one, land battles seem to be largely won or lost in the minds of the opposing commanders at certain key tactical and operational levels. The outcome may lie anywhere within a "long bracket" straddling all possible outcomes of the physical situation -- and even "impossible" ones like Goose Green.[5]

To focus once more on the blob in Fig. 4-3, one can imagine oneself as the attacker posing the question, "Can I make it grow and keep it going?"; then as the defender, muttering "Can I destroy or contain it, or is it going to run out of control and turn my whole position?" Note the link here with the classic phrase "turning a position". In a sense I am using the word "moment" as an abbreviation for "turning moment". And "turning" is the first key word I should choose to characterize Soviet tactics.

I think, too, one can identify the Anglo-American and the German-Soviet views in terms of main or "all-arms" forces -- the goose-eggs and big arrows, on the one hand, and the little blob on the other. The prophets of "position" are interested in destroying the big blobs and arrows, and see the little one as a nuisance -- a kind of mosquito that takes their eye off the ball. Attacking or defending, the apostles of "mobility" regard the fat arrows and the goose-eggs as a kind of parent matter that interacts to create, at one and the same time, the little blob <u>and</u> the possibilities of containing or destroying it. <u>Their</u> battle is about the little blob. So <u>we</u> have fighting machines that can move; and <u>they</u> have moving machines with at least a credible potential for fighting.

Or again, the prophets of position seem to see land warfare, whatever the tempo, as a game like chess with discrete, definable moves and pauses between the moves. We can speed this game up as they do on television, or slow it down by the kind of time-lapse photography biologists use to demonstrate plant growth. We can, I suppose, go on doing this until our eyes, even our brains, see a picture of continuous movement. But this <u>mental</u> sequence of alternating moves and pauses can never generate any kind of dynamic force as such.

I am sure the apostles of mobility, on the other hand, see mechanized operations as an uninterrupted flow -- now as slow as pitch, perhaps; now as fast as water. But always as continuous and as inexorable as music. It always strikes me very forcibly that the three leading proponents of armoured warfare -- the Germans, the Russians, and the Israelis -- are also the most musical peoples of our civilization. Then again, I suppose, one could argue that the Russians also excel at chess. And this brings me to a fascinating contradiction.

I have always associated armoured reconnaissance and light infantry with music such as the third movement of the Eroica; and an armoured advance with, say, the "Ride of the Valkyries" or the Grand March in "Aida". One looks at the map and sees the ground rolling toward one -- <u>se deroulant</u> as the French so nicely put it -- and the tanks rolling forward over it. In fact, one takes a bird's-eye view of a map, because that's the way maps are, but translates it mentally into a series of "low obliques". I'm fairly sure that's how the great exponents of the blitzkrieg saw it.

Not so the Russians. And I say "Russians" advisedly. For in the research I have been doing in the past fifteen months for a book, I have convinced at least myself that no officer trained under Communism has contributed a jot or tittle of creative thinking about the land battle. The Soviet concept of land operations has, of course, evolved and taken certain changes in direction; but it remains firmly rooted in the works of two Tsarist-trained officers -- Triandafillov,[6] who had the good fortune to die young; and the unfortunate Tukhachevskii,[7] purged by Stalin in 1936 and now being openly "rehabilitated" in military circles.

The Russian commander stays firmly with his birds-eye view. For him, to arrive is better than to travel. Both from leading authors like the ones I mentioned earlier, and from lower level articles in Voennyi Vestnik (a monthly very roughly analogous to Armor Magazine or Infantry but army- wide), I get a clear-cut impression of how the Soviet commander or chief of staff builds up his plan. As indicated in Fig. 4-4 he starts with a picture with the line of contact, or maybe a major water obstacle, at its centre.

He then looks at his primary operational objective, which could just be a key topographical feature such as the Rhine, but is much more likely to be the opposition's military nerve centres, their nuclear and other rocket artillery, and their operational reserves. For him, at all but the lowest levels, objectives are not pieces of real estate but enemy "forces and resources". And, by definition, the primary operational objective should be deep enough and important enough for its capture or destruction, as the case may be, to exert a strategic effect.

He will next look for intermediate operational objectives of the same kind on or near the most probable axis of advance to the primary one. Following this, he will look at the rear of the tactical enemy defence, at tactical reserves and tube and rocket artillery areas. This gives him his operational depth. Finally, he looks at the front of the defence and determines whether he can turn it or whether he will have to break in. You will see that he has established three foci: the primary operational objective; the area just beyond operational depth; and the centre of mass of the tactical defence.

He now balances these three foci by goose-eggs that will become his concentration area, the line of his "waiting areas" (in our terms, assembly areas), and the line of his deployment areas (not unlike our attack positions or forming-up-places (FUPs)) at which controlled movement of the leading elements will cease and they will break column. In the case of a river crossing, these will be the "preparation areas". The point I would like to make at this juncture is that, if we forget for a moment about tactical and operational levels, and say that the span of this picture either side of the centre line can be anything from fifteen to three hundred kilometres, we have a planning principle which works at every level from battalion to front. Start from the far end, work inwards, and balance. But let us now fix our minds on the three hundred kilometre scale, and front level.

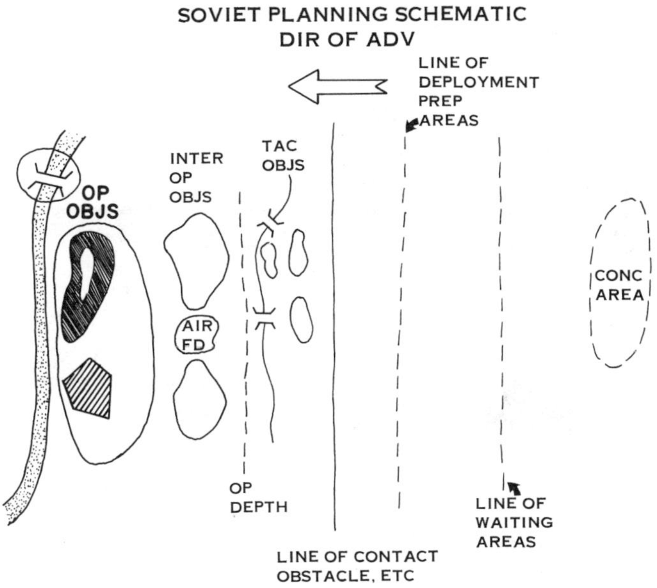

The Russian Commander's View
Fig. 4-4

Most will agree, I think, that, since this model will be superimposed on a map and will thus have a predetermined scale, it serves as a good pointer to the level of the operation and the "forces and resources" (or combat troops and supporting arms) needed; likewise to the echeloning and relief system to be applied; and again to the operation's duration and the logistic back-up needed.

Again (Fig. 4-5 refers) following Soviet practice and working from the far end, the front commander might order his tank army commander to find two operational manoeuvre groups, say one based on a tank division and one on the tank army's airborne assault brigade, the rotary wing formations now being deployed. The tank division would be reinforced by the number of raid forces required to peel off and deal with secondary operational objectives.

He would then look at the problem of getting through to operational depth, including the river crossing if there was one, and decide whether to deploy his two all-arms armies on a broad front, or to echelon them in depth. On present trends, the two armies would probably be deployed in depth, each in single echelon formation, with the tank army forming the front's third echelon. And so on.

I have deliberately gone through this process without basing it on a map to highlight the extraordinary love-hate relationship that exists between the Soviet military mind and ground. So far from regarding ground as a military resource, they appear to mistrust it. They consider it only in terms of avoiding bad ground; and almost every

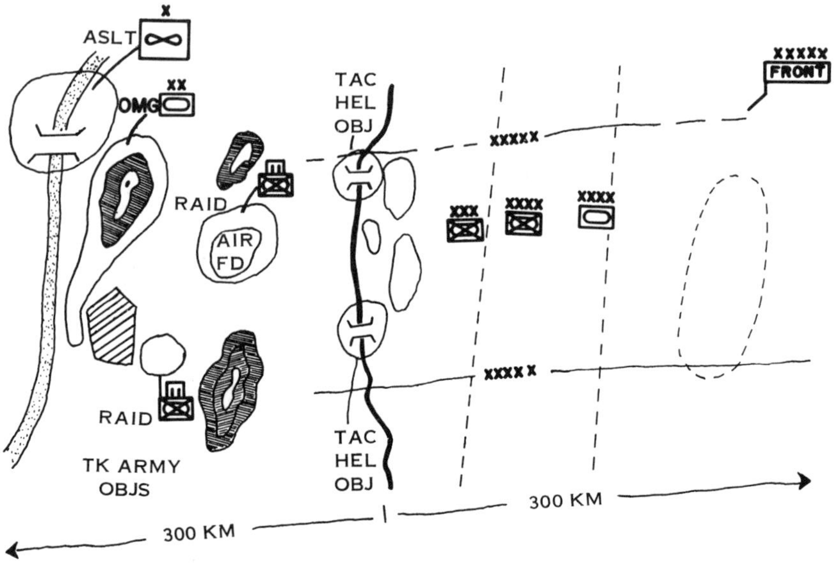

OMGs and Raid Forces in Concert
Fig. 4-5

type of terrain except the steppes is regarded as "special". This feature is, I think, associated with the Soviet interpretation of the doctrine of movement of masses rather than the concept itself. Certainly the Wehrmacht's great commanders were more aware than most of the value of ground, and of how to use it. And in the Bundeswehr the advocates of ultra-high mobility tanks and naval warfare on land have not got much of a hearing.

I hope, too, through the two approaches I have used, I have managed to stress the way that, for the apostles of mobility, the battle is all about that little blob in Fig. 4-3. For sometime the renaming of motor rifle formations as "all-arms" formations and the tremendous stress on the "all-arms battle" gave the impression of a deep-seated conflict in the Soviet Army between the tank men and the infanteers, with the BMP kings -- the "young Turks" with their swift, incisive slashes in the von Brauchitsch[8] tradition -- as a kind of joker in the pack. This no longer appears to be so. The all-arms force seems to be a mobile launching pad for the tank forces, while the newly deployed airborne assault brigades add, both literally and figuratively, a new dimension to operational mobility. For this reason, their combat worth is evidently equated in the Soviet mind with that of a tank division. By the same token, the changes in the balance of the tank division in favour of infantry and artillery appear to be linked partly to universal trends of military opinion, but mainly perhaps to acknowledgement of the need to fight without battlefield nuclear weapons.

To recapitulate, I am convinced that there are two clear-cut schools of thought about mechanized warfare -- each with its own variants, of course. Yet these two schools of thought seem able to co-exist, in fact, if we look at Leopard 2 and M1 Abrams, to produce two tanks as different as cheddar from cheddar. Should this analysis be along the right lines, it perhaps raises the following question: is the psychological content of land warfare in fact so high that two diametrically opposed tank philosophies -- or two radically different concepts of operations -- can, like real philosophies, live together in amicable conflict without having too much effect in the physical conduct of the battle, or on its outcome?

Before closing, I would like to extend my argument a little further in two directions. One is simply to suggest that the whole hypothesis I have erected and set before you is a house of cards, and that the name of the game is something quite different -- tactical, or perhaps better, operational, risk. I am not talking here about courage, either physical or moral. Readiness even to calculate high risks -- let alone to take them -- is something else again. Decisions to take the kind of risks the Soviets preach do not directly touch the physical courage of a commander high enough to take them. And, again particularly in the Soviet Army, it is usually the negative decision that calls for the greater moral courage.

To launch my final kite, the homework I have done for my books suggests that, given the rotary wing and fast tracked machines of the Nineties and a Warsaw Pact offensive based on the movement of masses, NATO needs either considerably more depth than it has east of the Rhine, or such powerful air-mechanized forces that, militarily, it should not be on the defensive at all. As was publicly acknowledged after Exercise "Crusader 80",[9] any form of conventional positional defence is going to get penetrated; on the North German Plain it could well be completely overrun. Quite apart from their cost, aggressive defences such as General Donn A. Starry's "airland battle", or the air-mechanized defence I proposed in my book Antitank, have the grave disadvantage that they resonate with the offence to amplify the dimensions of the battlefield, and to accelerate the tempo of the battle, even further.

A counter-thrust to the Oder or thereabouts, long favoured by some German tank men, is now being widely discussed in key British circles too. But it is very hard to see NATO agreeing in principle to pre-emptive infringement of the IGF by land forces, let alone authorising it if things come to a head. A counter-stroke after the event might not gain a worthwhile amount of room for manoeuvre; at best it would present the Soviets with the kind of encounter battle they seek.

The dynamism and tempo of the Soviet offensive, the limited size of the European theatre, the mounting dominance of firepower, and not least the strengthening trend of Western public opinion, all seem to point to some wholly different concept of defence. This could be aimed at containment, the trading off of the maximum of time for a minimum of space. Perhaps this is becoming a proper and sufficient aim of the defence at operational level, as it has long been at strategic and tactical levels.

Editors' Notes

1 Field Marshal Lord Carver, The Apostles of Mobility: The Theory and Practice of Armoured Warfare (London: Weidenfeld & Nicolson, 1979).

2 Harriet Fast Scott and William F. Scott, The Soviet Art of War: Doctrine, Strategy, and Tactics (Boulder: Westview, 1982), 224.

3 Viktor Suvorov, The Liberators (London: Hamish Hamilton, 1981). "Dnieper" was the first major manoeuvre to highlight a possible conventional campaign in Europe. The initially non-nuclear exercise, which encompassed a large tactical battle, airborne landings, and river crossings, was undertaken solely by Soviet forces but was attended by representatives of all Warsaw Pact armies.

4 O.A. Losik (ed.), The Buildup and Operational Employment of Soviet Tank Forces in World War II (1979) and A.Kh. Babadzhanyan (ed.) Tanks and Armoured Warfare (1970). Both in Russian. See Richard E. Simpkin, Antitank (Oxford: Brassey's, 1982). Losik was appointed head of the Malinovskii Tank Academy in 1969.

5 The battle for which Falklands Islands' settlement saw 450 paratroopers of the Second Battalion, The Parachute Regiment, attack and defeat 1600 defending Argentinian soldiers.

6 The Character of Operations of Modern Armies, written in 1929 by V.K. Triandafillov, called attention to the importance of the tank and considered it one of the most powerful weapons for any future war. He was concerned with the duration and depth of operations and the changes that tanks and other mobile weapons might bring to the battlefield. Harriet Fast Scott and William F. Scott, The Armed Forces of the USSR (Boulder: Westview, 1979), 72.

7 Marshal Mikhail N. Tuchachevskii, Chief of Staff of the Red Army from 1926 onward, was described by J.F.C. Fuller as a "remarkable general ... a barbarian who abhorred western civilization ... [and] had the soul of Genghis Khan .... Autocratic, superstitious, romantic, and ruthless, he loved the open plain lands and the thud of a thousand hoofs ...." He advocated highly mobile forces, which he visualized as a combined tank-air team, supplemented by parachutists and backed up by artillery and infantry. He stressed deep offensive operations, concentration of forces at the decisive point, and combined arms. On June 12, 1937, Stalin had this brilliant and impulsive officer shot, returning him, again in the words of Fuller, "to his god Annihilation." J.F.C. Fuller, Decisive Battles of the Western World, ed. John Terraine (London: Granada, 1981), 405-406.

8 Field Marshal Walther von Brauchitsch, a Brandenburger, was Commander-in-Chief of the German Army when Operation "Barbarossa" was launched. Relieved on request by Hitler in December, 1941, he had been an advocate of fast lightly armoured forces, based on mechanized infantry and operating in conjunction with tactical air, as the primary strike force. Simpkin, Mechanized Infantry (Oxford: Brassey's, 1980), 14.

9 Exercise "Crusader 80" tested the lines of communications from the UK to I(British) Corps; 10,000 regular army and RAF and 20,000 Territorial Army personnel, together with about 4,000 vehicles and a small amount of equipment, were successfully moved by air and sea from the UK to the Continent.

CHAPTER FIVE: THE ASCENDANCY OF FIREPOWER

by

Professor D. S. Graham, MC
Centre for Conflict Studies
University of New Brunswick
Fredericton, New Brunswick

The title of my presentation, "The Ascendancy of Firepower", reflects a rather controversial, even pretentious, tactical military theme and I had perhaps better explain what I mean by it. I refer to the initiative that increasingly accurate weapons with higher and higher rates of fire and more lethal projectiles attained over things that simply moved. Now that's not the same thing as saying that defence gained the ascendancy over attack, nor that what was true then -- and I'm talking about the period 1904 to 1945 -- is necessarily true today. I don't want in any way to give the impression that my analysis, that my research in this period, is necessarily going to be useful to you today. If you find it so, then I shall be glad, but I won't try to pretend that I am uncovering universal truths; I'm just an historian trying to record what I've found and perhaps leavening it with some of my own skepticism. I'm not going to offer a narrative of the events of the first half of the century which is what Shelford Bidwell and I attempted to do in our book, Fire-Power. What I am going to try to do is present some of our analysis, which, had we published it in the book, would simply have bored most readers to death and the book would not have sold.

As professional military students, we have all been brought up on the conception of firepower and movement, employing one expression after another over the years to represent the two elements that have often been reciprocals. These have all been used in the manuals, which by the end of the nineteenth century generally tended to focus on the terms fire and movement. The soldiers' problem though, is to practise the general ideas that are in the manuals, for the manuals, themselves, are written in very general terms. There is a gap between their theory and soldiers' practice.

Sir James Edmonds[1], talking back in the period about 1910-1912, had a lot to say about them. In fact, though he had something to do with the writing of them, he never accorded them much respect. He claims that they were scissors and paste jobs, constructed by men who plagiarized and knew little about the art of war in practice. He described most of the British manuals of the years 1910 to 1914 as plagiarized from the German, translated into passable English, and then simply regurgitated. They were neither analytical nor handyman's guides to tactics like the German barrack room publications. An Australian remarked that they were about as much use to his soldiers as cuneiform inscriptions on a Babylonian brick. A platoon commander of the same period said that manuals were useful to the point that you entered the fire zone, then they told you that you were on your own. In the last 300 yards of attack you had to close with the enemy without being destroyed, though in defence you had to stop the latter cold. That much information they gave you, but how you did it was really up to you. Whether it was the last 300 yards or the last 1200 yards, of course, varied with the ground and the weapons, but sooner or later men and vehicles had to face the fire storm. The quicker they closed on their objective, the fewer the casualties. That seemed to be the simple message of manuals in 1914.

It was, as I'm sure all of you realize, an over-simplification of the infantryman's problem. Men and vehicles are so often cut down by fire from positions not under attack and from positions not previously seen. One of the most difficult things in war, in my experience, is actually to see the enemy. The battlefield seems to be empty except for smoke and things that fly. In defence, your friends in other positions protect you best. As a result of the marked increase in firepower between the last years of the Nineteenth century and the opening of the Great War in 1914, a crisis in tactics crept up on the armies which I think has been with us ever since. Solutions to the crisis were proposed before 1914, many of which were tested in the war that followed; others were tried in the Second World War. While some of the latter attempts could certainly be compared to reinventing the wheel, progress was nevertheless made, particularly during the period 1942 to 1945. But many tactical solutions are but improvisations. For, as the values of the parameters that determine the price to be paid by moving things in the face of firepower are variable, there can, in my opinion, be no finality, no final solutions. All we can do is to study what has happened in the past and hope that we can be wiser another time.

Tactics are, to my mind, really about this last few hundred yards. All enemies are very much the same until they are locked with you in the fire zone; there you can see what they are made of and they can see what you are made of. It is quite easy to be relatively brave 800 to 1000 yards away from the enemy who is firing at you -- I mean, relatively brave -- but when you get close, it becomes a question of who can stare who down. Now I do not need to explain to an audience like this the characteristics of direct, indirect and predicted fire, or rifle and automatic fire; obviously you know about these things. But what I want to talk about today in my analysis is the interdependence of familiar tactical elements in what was, in the period I'm talking about, a fairly dynamic process. I think that the inter-relationships that developed may be a little less familiar to you so my analysis of the story up until 1945 will focus on these.

First, there are relationships between the arms[2] that you might not have considered exploring in your study of tactics. These are functions of the institutional characteristics of the arms. I have been fascinated by trying to understand what makes a gunner tick, an infantryman tick. What is the difference between the insides of their heads? This interests me and so I thought I'd start off by recounting some thoughts that I have about the way the different arms approach tactical problems and how, therefore, it may affect the way they solve problems. Obviously the arms I have to talk about, if I'm talking about the earlier period, are infantry, cavalry, armour, and from 1916 onwards, artillery, and engineers.

The first thought that occurred to me was that in the first half of the century, the period under review, these arms each manifested certain characteristics to a greater extent than others. The infantry, cavalry and then armour -- and I admit this assessment is probably controversial -- tended toward "purity of race", by which term I am sure you will know precisely what is meant; they sought simplicity of aim and of organization. These arms habitually behaved like consumers. The second group, the artillery and the engineers, on the other hand, provided a service that made them complex, flexible and subordinate; they were consequently problem solvers, students of the market place that they served. That is the first point I would like to make, in particular, the business of subordination.

My second point is that arms behaved like political and social institutions, especially in peacetime. They made proprietary claims to weapons and functions that, in wartime, obstructed problem solving. Let me give you an example: the Royal Artillery was still divided in 1914 into two distinct parts, the Garrison branch and the Field branch. I would have said that in Field the sword was mightier than the pen and in the Garrison the pen was mightier than the sword. In other words, the Garrison tended toward scientific methods, whereas the Field imagined artillery support as galloping into action with the glint of steel and the whites of the horses' eyes showing. It was essentially, though unfortunately not only, a social and intellectual division; it was a political division too. If you passed out of the "Shop" (the Royal Military Academy, Woolwich) low down on the list (even in 1917) you went to Garrison; you had to do better if you wanted to go to Field. This was where you met the rising stars who were in the expeditionary force divisions. If you wanted to go places you went to Field. Yet it was a much less brainy form of artillery than the Garrison and when the war came it was the Garrison brains that were needed. By that time, however, the Garrison gunners were not in a good strategic position in the army; their career structures were not well-developed and we consequently suffered for it because it took some time before their ideas were heard in the right places.

A further example that you will find more controversial has to do with the proper role of each arm in the tactical seizure of ground. When a battle was aimed at seizing ground the arm responsible for that task was placed in the driver's seat. Equality in making the plan was unlikely then, even if it was desirable. The arm that had to seize the ground was considered the boss, the action of seizing ground itself becoming a sort of "boss performance" for a period. Let me illustrate at this point how it is supposed to have worked according to some historians and how it actually did. For a period on the Western Front, the artillery was expected to conquer by destruction and the infantry to simply occupy ground. So many of the history books will tell you it was an artillery war -- but the artillery was not at all in the driver's seat. The people who made the plan and more or less decided what the artillery ought to do were

An Arm in Eclipse
Fig. 5-1

the infantry, because they actually occupied the ground. The whole performance was done for the infantry's sake; it wasn't done for the artillery's sake at all. In sum, it was an artillery war, but with the infantry still in the driver's seat because they were the people who had to take the ground. In the original tank corps plan for the Cambrai attack, for another example, ground was not to be seized; the mode was essentially movement and disruption. In that plan the tankers would have been in the driver's seat. Finding this quite unacceptable, the infantry stepped in and turned the plan into the seizure of ground; they thus became the bosses of the plan. It failed because the ground could not be held, and in one battle -- Flesquières -- we all know what happened when the 51st (Highland) Division hung back and allowed the tanks to be shot to bits.[3] It is an example everyone should remember. It was the result of having one arm, the infantry, in the driver's seat going its own way regardless of the needs of another.

Seizing ground is the result of movement. Movement is made possible by fire. What, though, is the object of seizing ground? It is indeed remarkable that in this period, particularly between 1904 and 1914, no one actually asked that question, or if they did, they certainly didn't get a very sensible answer. I would argue that the object of seizing ground is to seize new fire positions and new observation from which fire may be directed, or to deny those facilities to the enemy, although I will agree you can accomplish that in other ways. By accepting this as the object of seizing ground, you have the link between fire and movement, the common aim of the fire element and the movement element. A virtual monopoly of firepower in the hands of one armed institution, however, and a similar monopoly over movement -- interpreted simply as seizing a piece of real estate -- in the hands of another, makes it extremely difficult to strike a balance between arms and between these two tactical elements. Institutional behaviour, in fact, is likely to impede a solution.

With the appearance in 1916 of a new arms institution, namely a tank corps that reflected the introduction of a new weapon, it looked for a while that a bridge, both institutional and tactical, had been created. Unfortunately, the tank's tactical nature was not defined and the army as a whole was reluctant to allow another institution with the proprietory rights to new weapons systems to come into existence. The mold, as it were, could not be broken. The new corps was not going to be allowed to move, seize ground, generate firepower and behave like traditional cavalry as well. As a result, I don't think anyone was really clear about what the new arm was supposed to do and the institutional enemies of new arms were out to make certain that it was going to conform to their preconceptions. The infantry, the seizer of ground, was to remain "queen of battles". The artillery was to remain the generator of firepower.

My final reflection on the effect of institutional thinking on fire tactics, which is really almost an aside, is that the debatable land between the seizer of ground and the generator of firepower has been much fought over and it is still rather a no-man's-land today. Given the dominance of institutional thinking in peacetime and the fact that it takes a great deal of shaking in wartime before hard tactical priorities replace habitual ones, it is really no surprise to find that effective action is left in the hands of individuals. One can't expect institutions to move quickly or positively. <u>It is individuals who change things</u> and usually the individuals are those who can play the system if they're conservatives or overturn it if they're progressives. In either case, since individuals tend to be subjective, cold logic and hard reasoning have rarely prevailed in the history of tactical development. It remains, in fact, a story about individuals and obstructive institutions and the brute force of weapon systems. I do not believe, of course, that the cause of military enlightenment is helped by labelling actors, whether they're institutions or individuals, as black hats and white hats. So please do not think that because I am a gunner, I am therefore claiming that gunners are wearing the white hats. This isn't my belief at all; in fact, I've already pointed out in discussion of the Garrison-Field business that I rather feel the artillery wore black hats on that particular occasion. I do think, however, that we should understand these human as well as technical factors. The tactical analysis that follows will throw light on how things actually worked. I am not suggesting this is how they will work in the future or how they do work now.

I would like now to describe the position of one arm after another in certain situations at certain times to define the debates that actually occurred in the period 1910-1914 and to show what happened. I shall start with infantry, the master arm, no doubt about it, on which debates about firepower and its relationship with movement were centered.

Infantry takes ground and holds it and, as many of the manuals said and most of the senior officers repeated, taking ground was advancing and advancing was winning; therefore, the infantry wins wars. The debate going on at infantry battalion level at this stage was whether to reduce the man:firepower ratio or increase the firepower:manpower ratio. What it really boiled down to was whether a battalion commander was prepared to have fewer men in the assault for the benefit of more supporting firepower. One can imagine the politics entering into the replies: "I'd like both", or, "Over my dead body will I have fewer men in the assault".[4] At this particular time, of course, the effectiveness of the firing line which could obtain fire superiority over the immediate enemy was simply measured by the number of rifles in it. You built up a firing line, which meant that you did it incrementally, by the adding of people. This meant, in turn, that you directly related manpower with firepower, which we would now tend to regard as a mistake because the general tendency since has been to think of manpower and firepower as reciprocals rather than directly related. By 1913, the infantry had taken one step in recognizing this. They had introduced the new small section which was half the size of the old section; instead of 32 there were now 64 of them in the infantry battalion. Those who had supported the change recognized that the new rifle would increase the firepower of the section and that, by and large, the section of 10 men was really a sufficient firepower unit for the time and about all that one man could control. However, the overall number of men in a battalion remained virtually the same and so did its weapons system. So it is doubtful whether the introduction of the new section indicated any fundamental change in opinion about the manpower:firepower ratio.

A dislike of "specialists" in the battalion influenced the debate about manpower versus firepower. You already had the machine gun section of two guns which was handed out to a subaltern, always a subaltern, but sometimes to quite a newly arrived one, with no political power in the battalion at all. He was then given all the toughest soldiers with long records of misconduct and, usually, a sergeant who was able both to keep the subaltern straight and the men in line. Obviously, if the machine guns had been brigaded into batteries of eight a captain would have commanded. Since that rank has just that much more political clout than a subaltern, the automatic firepower lobby might thereby have been strengthened. This didn't happen because the people who ordain establishments were just as bloody-minded then as they are now; they would not give an extra captain for the brigade and no battalion commander wanted to give one up. The machine guns, therefore, remained with the subalterns who had no influence, with the result that this particular breed of specialists did not exactly flourish in the pre-war army.

There was, however, another consideration in this regard. If you have a lot of specialists in a battalion, this, generally speaking, cuts down on that abstract relationship called homogeneity and homogeneity of weapons systems leads to cohesiveness. It has generally been found that stress is higher and morale lower in heterogeneous units; stress is also higher and morale lower amongst people who work at separate jobs in the battalion. If you have a lot of odds and sods doing different fire jobs in the battalion, there is a danger that the cohesion of the battalion may be lost. I believe a lot of people felt that it was a bad thing to have increments of

The Real "Queens of Battle"
Fig. 5-2

firepower of any sort if it meant that you had to have special units. Much better to stick to the good old rifle companies which were homogeneous.

The staff also looked at precisely the same problem and, as they always do, viewed it from a slightly different angle. The staff had no arms schools at this time to advise them; therefore, most of the control of training and doctrine was in the hands of the single corps commander that existed at that time, divisional commanders, and battalion commanders. There were no schools to inject forward thinking and so on. The debate from the staff point of view thus came to centre on the question of whether battles were decided by firepower or by the infantry advances that firepower simply enabled. Here the School of Musketry produced some evidence that would suggest that the firepower of the infantry battalion in the defence was perfectly adequate but that it needed more firepower in the attack. The defence was very powerful and pretty fearsome, but more light machine guns and automatic rifles were required in the attack. The staff, though, was afraid of the idea of increasing firepower in the attack in the infantry battalion because it would encourage infantry battalions to have fire fights with its opposition -- to "play at long bowls" with the enemy. It might encourage the idea that you could shoot the enemy out of his positions and the staff was convinced that you never could do that. When would such a fire fight end? In Manchuria, 20,000 rounds had been fired for every casualty inflicted on the Russians by the Japanese. This was a little lesson you could not ignore so maybe it was a mistake to encourage more automatic weapons. There was also the fear of reducing assault manpower and the generalized objection to heterogeneity in the army through developing variety. All sorts of problems arose if you did that.

The staff, in fact, arrived at a compromise which was to hand the problem over to the artillery. They told them to use their guns like the French used their 75mm gun which was a high precision, light, high muzzle velocity, very flat trajectory weapon. The Royal Artillery was, as far as possible, to engage the enemy over open sights or at least from semi-concealed positions and give direct support to the infantry. That way the infantry would have its firepower increments. The infantry, for its part, was told to emphasize gutsy performance, high morale and rapid movement. The staff solution, then, was to imitate the French and to have the artillery use French methods to support the firepower of the infantry.

Let us just have a quick view ahead at this stage to see what, in fact, really happened when the war came. The immediate discovery was that the infantry did not have enough integral firepower, even in defence, because close country and casualties to men and weapons meant that the two machine guns were not adequate. And gradually, as the war went on, a variety of other support weapons was introduced into battalions and brigades. They crept forward, notably light machine guns, from being battalion weapons to company and platoon weapons until eventually they were section weapons. The platoon had four sections, as it had in 1914 but with light machine guns in two of them. So the discovery, of course, was that more, and a greater variety of, integral firepower was needed in the infantry battalion.

I would now like to look at the firepower debate from the point of view of the artillery.

The problem for the artillery was to work out the implications of the debate about whether firepower was decisive or "enabling", which the staff had decided in favour of the latter. The debate, inside the artillery, was over how you gave efficient direct support to an infantry battalion in the attack whose integral fire support was insufficient? The need to do that was quite recognized by the gunners who understood the infantry problems. The gunners, of course, turned the matter into an artillery problem. Every arm tends to do this, for though it is the same problem, it somehow looks different if you wear a different coloured hat. In artillery terms the question was whether neutralization or destruction was the purpose of their fire. Neutralization meant certain things, certain sorts of weapons, it meant high rates of fire, a degree of accuracy, surprise above all, direct fire if possible, and close targets and forward guns on forward slopes which had been shown in Manchuria to be disastrous. The artillery tried to find a way of supporting the infantry, particularly over the last 300 yards. In peacetime, of course, this could not be practised because of the serious risk to infantry. In any case, it meant that the artillery had to get into positions from which it could lend close support to infantry. It had to be almost with the infantry, its guns virtually amongst them; otherwise, how could you find out where they were or what they were doing? This tended to lead the artillery forward into dangerous positions, the thing that they had learned from observation in Manchuria to avoid. In Manchuria reverse slope positions were the only way in which the artillery could survive and now they were being asked to go forward and to perform right up front. The alternative to this, of course, was to go for destruction, not neutralization, and destruction meant, of course, heavier weapons, more ammunition -- much more than they had in peacetime -- indirect fire by and large, and even prediction. It meant that you had to have much better communications, particularly lots of cable and good telephones, for occupying reverse slopes meant that you had to have your guns further back which, in turn, called for more telephone line. If you did use indirect methods you could get more guns onto each target. If you have all your guns in the forward area, the forward area gets very crowded. If you have them further back, you can actually deploy more guns and use them to engage the target that the infantry is attacking. Longer ranged artillery is an advantage under these circumstances. What you have here, in fact, are two fairly venerable choices that the artillery has been grappling with and probably still does grapple with -- neutralization versus destruction, and the relative advantages of direct fire immediately applied over heavier fire which takes longer to arrive and may not be so accurate.

Soviet Artillery Still Aims at Destruction
Fig. 5-3

What the infantry and staff didn't grasp was that the artillery at that particular time had learnt from Manchuria to take the road that the Germans had taken. Its weapons system array was complex. It comprized howitzers, field guns, and heavier calibre guns. It was trained to use destructive methods if necessary. It also sought to improve its communications so that it could give fire support by whichever method was appropriate -- direct or indirect. This choice had been taken away. Gunners were told to adopt a French philosophy with which, frankly, they didn't agree because they'd seen it in practice on the ranges in France and they weren't terribly impressed with the accuracy of French gunnery. They were nonetheless told that because of the weakness of infantry firepower, they would, by and large, have to use these direct methods for which their weapons were less suited than were those of the French.

Now, let's have a quick look at the view ahead at this stage. The inter-locking nature of the infantry and artillery arms had been addressed but not satisfactorily resolved by 1914. The staff started with a premise of the infantry's primacy because of its ground-taking and ground-holding role. The methods and weaponry of the two arms were not integrated because the debate about the purpose of holding and taking ground had not been settled. When the war actually came, infantry firepower advocates -- the people in the infantry battalions who, frustrated by the weakness in the integral fire support, had said "Let's have more light machine guns" -- leaned on artillery destructive fire more and more and thus lost their grip on fire tactics. They became addicted to artillery because of their own lack of integral fire support and when they got it, not having practised it in peacetime, they didn't understand how to use it in war. They, therefore, had to learn the hard way.

By the time the war came, the artillery could not answer the question: "When the companies disappear into the smoke, who shall save them? -- themselves they cannot save." Though the story of the subsequent development of scientific gunnery to support the infantry is quite a long one, I shall attempt to give it to you very briefly. In the opening year of the Great War, the whole of the Western Front had to be completely remapped. This activity was marked among other things by the introduction of aerial photography, surveyers dashing about with theodolites, the rediscovery from Garrison of how to calibrate guns, the art of taking the temperature of the charges, the introduction of trig points and finally of bearing pickets. By the middle of 1917 the artillery could come close to hitting a target without registering it, providing it knew where the target was on the map and where the gun was on the same grid. This, in turn, allowed the artillery to achieve surprise because long registrations, long dumping plans, long preliminary bombardments were unnecessary. These had previously given the show away before the infantry advanced. This development took two years having started in 1915 with survey and continued until the time of Cambrai, in November 1917, the first battle in which all these methods were actually used.

To jump ahead a bit, I would say it was not until 1942 that we finally reached the point where we recovered the scientific gunnery mastered in 1917 and 1918. With the addition of the radio, the artillery was able to give the infantry in the smoke some reasonable support; but only when the radio worked, for even with the "walkie-talkie" and rebroadcast "19 Sets" in 1944, the close support of infantry by artillery forward observation officers (FOOs) remained a chancy business. So with scientific gunnery, you could start a battle off with a great bang and achieve great surprise but when the infantry disappeared into the smoke, you still somehow had to <u>contrive</u> to give them that close impromptu support. Despite the advantage that the artillery had given the army in pre-arranged, pre-set battles, this gap remained between the level of the fire programme and the beginning of impromptu, low level support. In the impromptu battle, the gunners still could not manage to give assured support to the infantry. The infantry consequently realized the need for its own integral fire units.

To return now to the infantry, and to what it was doing, I like to feel that it made a deliberate effort to get away from the artillery crutch, but I'm not sure that it did. Its actions were much more ad hoc. First, in the original trenches of 1915, all sorts of specialists began to appear and the infantry battalion lost the homogeneity that it had had in 1914. It acquired the Stokes mortar in 1916, a very unpopular brigade weapon. "Will you take my punishment or go to trench mortars?" was the sort of sentence delivered in a company office at the time. As for the artillery mortars, which I do not believe were any better managed, who on earth would want to "give up his horse for a piece of drain pipe?" This was the prevalent artillery view of the mortars. But there were also the people who threw grenades of one sort or another, "the bombers". They became an elite and very specialist force, very useful in close-trench fighting. As soon as semi-open fighting occurred, as it did during the Somme and afterwards, the bombers were less useful. They ceased to be "specialists" organized in battalion units and became part of the platoon fire team. The light machine guns began to seep forward starting as company weapons; by 1918 they were in platoons. The rifle grenadier gave a platoon its own light artillery that could reach out beyond the distance a man could throw. A variety of grenades for different purposes was provided, All this meant that there was more integral fire support in a platoon and that platoon and section commanders had more responsibility. The later stages of the Somme marked the beginning of this trend, which has continued ever since.

In the meantime, the section of medium machine guns in each battalion had been removed from the battalion to become a divisional weapon. Here we have a rather sad story, a little less sad in the Canadian case than in the British. A new corps was formed to look after the machine guns, and new corps in the British army generally tend to be unpopular. Unfortunately, this was a corps run by people who were prickly and pushy to the extent they got everyone's backs up. Because they themselves were difficult characters and ambitious, it was assumed that the machine guns were an ambitious weapon in some mysterious way and ought, if possible, to be thwarted. The machine gunners could see precisely what they should be doing. They were to try to fill in the gap I mentioned earlier, the gap between the artillery weapon and the infantry integral weapon. Regrettably, while they did fill it to some extent, they didn't really do it as effectively as they ought to have. One of the reasons perhaps was that so many of the machine gunners were not really true infantrymen. This was the view of a German observer and also of a very astute British observer, H.C.C. Uniacke, who was at that time a Major-General Royal Artillery (MGRA).[5] He felt that the machine gunners were fascinated by the mechanics of their art, that they were trying to be gunners. In attempting to master barrages and indirect fire they were not actually using ground and getting up with the infantry to give them that close support that they badly needed. By and large, I do not think the British machine guns in the Great War were a success. The Canadians, allowed half as many machine guns again as the British divisions (96 as opposed to 64), were more successful in this respect, perhaps because they were not so obsessed with arms institutionalism.

There was, thus, a gradual diversification of weaponry in the infantry battalion. None of the new weapons was really mastered, however, because none had been in use before the war. The solution to the close support gap was thus perceived by many people to be the tank when it arrived on the battlefield in September, 1916. In my opinion, the infantry was not ready for the tank. The artillery had developed a new form of neutralization to help the infantry on the Somme; the new form was the creeping barrage and it was marvelously successful. Unfortunately, it was linked with linear tactics rather than columnar tactics, which meant, in fact, that the infantry tended to simply advance in lines behind the linear concentrations of shells. As everyone knows, it is very difficult to generate efficient firepower from moving lines, which are also hard to manoeuvre, so the very success of this neutralization method hamstrung infantry fire tactics. For tanks to be really effective they needed to work with small groups, with section columns, as opposed to company lines. The infantry, in short, was not ready for the tank when it finally arrived.

The question of the tank, whether it was fish, fowl or good red herring, remained unanswered through the Twenties and on into the Forties. The question was nonetheless asked because the tank, contrary to expectations of the time and retrospective opinion, <u>was not obviously the decisive weapon</u> of 1917 and 1918. Although marvelously quickly produced, it was incredibly unreliable mechanically and a fiendish thing in which to fight. Its noisy engine gave off extreme heat and noxious gases. Machine guns firing out through the ports produced additional noise and gas. Poison gas being sucked into the chamber often forced the crew to wear respirators. Temperatures rose to 105 degrees. After fighting in these conditions for 24 hours crews were not effective. The tank, because of its undeveloped mechanical nature, apart from other factors -- employment and crew training -- was, therefore, not a huge success.

I do think, however, that the tank's lack of success was due as much to faulty employment as to mechanical conditions. Tankers themselves, even after the war, were not quite sure and spoke with different voices about what the role of the tank was supposed to be. One should also remember that when the role of the tank was being discussed after the war, the majority of infantry officers had never fought with tanks. Even certain tank battalion commanders had never commanded or even served with tanks for they had been in the infantry or some other arm such as machine guns. In the 1920s there wasn't a well-documented body of knowledge about what the tank had done in the Great War; therefore, not very much straight thinking was done about what it ought to do in the future based on experience in battle in the past. It is indeed extraordinary that no official analysis of the Great War operations was undertaken until the Kirk Committee Report in 1932. That was an awfully long time after the war and time enough for all sorts of prejudices to develop.[6]

Abrams M1: With Crew Comfort and Continuous Battle in Mind
Fig. 5-4

As for the experience of the Great War, the first point is that the tank was conceived in the conditions of late 1914 and early 1915. These no longer existed by the time the tank came into use in September of 1916. It was originally designed as an infantry support weapon to go across a fire-swept no man's land 200-600 yards wide to grind down barbed wire and possibly machine guns as well. By the time it was used in 1916, semi-open conditions really demanded an entirely different sort of weapon with a different sort of conception. It needed to use its guns more and to work with small groups of infantry fighting their way through a defence in depth. It had to have very much more endurance and range and its crew had to have very much more tactical sense.

There were by then, of course, two more conceptions of what the tank was supposed to be. One of them was the conception of Fuller (in my opinion not one of his happier ideas) that the tank was just "an armoured horse". It begged, in my opinion, an important question: should and could the tank seize and hold ground? It was a question that the cavalry had already been forced to face up to before the war, when it was firmly told it would have to seize and hold ground, even though not for long, despite its fear of being turned into mounted infantry. Should the tanks be teamed up with specially trained infantry to enable them to hold ground as a team? The German cavalry, having refused to hold ground, had been teamed up with jaeger battalions that were given more machine guns than an infantry battalion, and mounted in vehicles (these were, I suppose, the first panzer grenadiers). Then there was the question of whether tanks should fight their way through a defended area like the infantry, or were they, like the cavalry, simply waiting for the "G" in gap? In my opinion, Fuller's answer to the first question was "no", even though it may have been perhaps an equivocal "no" in some places. I believe, however, that he answered "no" because he had found that cooperation with other arms was so difficult that working alone was much to be preferred. Cooperation was not of course Fuller's own strong point, but he had something to complain about in 1917 and 1918.

A third view of the tank was derived from the special conditions that existed in late 1916 and in the 1917 and 1918 offensives. It was that the tank should fill the place vacated in 1915 by the artillery and not adequately filled by the machine gun corps or by any other weapon: namely, to provide direct gun support to the infantry or to other tanks in the dog fight as the infantry or the tanks fought their way through deep defended areas.

Diversification in tank design and function was as inevitable as was conflict between upholders of each of the three conceived tank roles. Lack of analysis in depth of the fighting in 1917 and 1918 led to too many untested hypotheses and eventually to a garrison state mentality in the new tank institution, which it justified by pointing to the antagonism of older arms. Lacking scientific methods and objective analysis it never arrived at honest conclusions. Hence, features of the tank arm in 1939 were mechanical unreliability, pure-in-race ideas, an armoured division short of infantry, the scrapping and neglect of assault guns like the Birch gun, and an ignorance about the cooperation of anti-tank guns and tanks that prevailed as late as 1942.[7] Had the lessons of the Great War been properly studied on the other hand, and arms institutional strife controlled, the story might have been happier. I would conclude that the tank arm never rose above the provenance of its officers in the other arms. They brought with them no clear doctrine about cooperative fire-tactics.

I do not wish to leave you with the idea, however, that I believe cooperative fire tactics were not achieved between 1942 and 1945. It is my impression that German formations worked their tanks, grenadiers, anti-tank guns, and assault guns together extremely well, although I know that there were some German officers who complained bitterly at the time. But then German military writing was incredibly pessimistic and they were always complaining how awful everything was and how badly they were doing, even while they were obviously winning or on the verge of winning the battle. I think it was their professional sense of how perfect everything might be that made them write in this way. We were much more used to muddling through, so we accepted second best. But I think that the Germans were most successful precisely at the level of which the British were often weakest -- the troop/company/section of guns. The Germans believed in the direct fire of high explosive (HE) weapons in 1940-45, just as we had before 1914. They often turned small battles to their advantage by the initiative of small groupings. Now this has always suggested to me that they grasped the reason for seizing ground: to obtain advantageous fire positions. It was a progressive and often impromptu procedure. Leverage describes it, and it seemed to work according to principles which were commonly understood.

I believe that General Alexander was struck by German small-unit actions, and when he got back from Dunkirk, he realized that the British infantry, particularly, was lacking in instinctive battle drills.[8] They didn't know what to do when they were faced with a variety of situations. There was no instinctive summing up of situations and then rapid action, and I think battle drill was General Alexander's attempt to remedy this weakness. Battle drill started, I suppose, in the rehearsals for battles in 1916-18. In 1941, it began to gather way and by 1942-43 seems to have been fully developed at Barnard Castle. But it only worked in the field in 1944, when young officers brought up on nothing else had their chance. I suppose Alexander was suggesting that the initiative of the junior NCO and junior officer in the German army was not just an abstract quality, but the result of learning instinctively which of a number of set drills ought to be adopted in any one of a number of tactical situations.

The origins of battle drill are at least as old as the 18th century when battle drill and parade ground drill were one and the same thing. At the Canadian Infantry School in Vernon, British Columbia, in 1943, battle drill was practised on the parade ground first. An American observer who was at Vernon at the time, said that that was the main difference between Canadian battle drill and American; the American didn't do it on the parade ground but went straight into the field. It strikes me, however, and I may be wrong here, that in Northwest Europe during the Second World War, the excellent idea of battle drill was shown to be rather skin deep. When casualties were replaced by new men and when more and more Great War methods were used such as enormous bombardments and long creeping barrages, great walls of chinagraph denied battalions the room to manoeuvre. These linear advances replaced what the Australians had called, in the Great War, "peaceful penetrations" by sections and platoons in columns. Certainly I found in some battles in Normandy that I might just as well have been on the Western Front. I can remember in 1945, in Operation "Veritable"[9] (a rather ghastly blood-bath where the 5th Coldstream had over 200 casualties), seeing those lines walk forward, my own FOOs with them. Really, it was like the Somme. We gained the objective but at a high price.

A shortage of trained "rifle" infantry within complex battalions, and of infantry within complex infantry divisions, evident before June 1944, even, brought into question in the Second War, as in the First, the wisdom of having such a variety of special units in the army and an examination of the best balance between the arms. This old question came up in 1917, was examined by the General Staff in 1919, and again in 1944-45. Whereas it was argued in the winter of 1917-18 that "we can't afford all these 'funnies', all these tanks and machine guns and things; it's the infantry that really wins battles and we mustn't deprive them of men," in the Second World War we really went the other way until the end -- although we turned anti-aircraft units into infantry. We had a plethora of glamorous armoured brigades for instance, while the infantry battalions supporting each of these were was down to three companies much of the time. It was infantry that we were so short of in the end. So things came full circle. We had gone overboard producing masses of firepower and we didn't have enough people to actually walk forward. This reflection, of course, makes one more sympathetic to the conservative arguments in 1914. On the other hand, maybe we were using Great War methods in 1944 and that's why we were short of infantry. But short of it we were.

In conclusion, it seems to me that in both wars our methods were objectionable on the grounds that they were too reliant on materiel and time. Normandy and the Rhine battles were expensive both in materiel and time; they were very "pompous" battles. In peacetime, economy usually suggests that mobility rather than firepower should be pursued; in war it is always the other way about. In my opinion, the necessary connection between the two should be pursued, rather than one or the other. A new factor today seems to be endurance: for how long can sophisticated equipment function? Can it be replaced? Are reinforcements capable of operating it? We might also ask how flexible is our methodology and can we return to first principles if necessary? Can we communicate reliably in all circumstances? How effective is our target finding facility, without which firepower cannot be ascendant? What is the rationale of the battle of the future, is it to destroy the enemy regardless of where it is done? Is taking and holding ground rational, and if so, in what circumstances and for what purpose? Do we today have a better grasp of the proper use of firepower than we had before either of the two wars?

Notes

1 In the years 1910-1914, Edmonds was a general staff officer. After the war he became the official historian of the British Army's campaigns.

2 This word "arms" is British shorthand for fighting elements in contrast to "services" such as transport and ordnance.

Editors' Notes

3 The commander of 51st (Highland) Division, Major-General G.M. ("Uncle") Harper, an infantryman of the old school, refused to have his troops rehearse the new tank battle drill that had been worked out by Fuller. Instead of having his tanks advance in sections of three each in prescribed arrow-head formation with infantry in file right behind them, Harper chose to have them move in groups of four in extended line with the infantry in similar line formation 100 yards behind. Whereas battle drill also laid down that tanks were to turn left after crossing trenches, Harper obstinately directed them to turn right (left or right did not make any difference, but it was obviously preferable that all tanks should go in the same direction). The result was that Harper's infantry had ultimately to search to find the gaps made by his tanks and they eventually fell behind by up to 400 yards. German field artillery subsequently engaged Harper's tanks with direct fire and had a devastating effect. It was argued that had the infantry been close behind the tanks, as Fuller had planned, these guns could have been silenced. As it was, the tanks were left on their own and 27 were knocked out in the first few hours, the only hold-up in the entire Cambrai attack. Ironically, Fuller had originally suggested that the Flesquières position not be assaulted frontally but rather outflanked by tanks. This was not conceded. What was conceived as a raid operation thus became a struggle for ridgelines, one of which was

Flesquières. Bryan Cooper, The Ironclads of Cambrai (London: Souvenir Press, 1967), 64-68, 79-80, 117, and 222-223; and Robert Woollcombe, The First Tank Battle: Cambrai 1917 (London: Arthur Barker, 1967), 35, 44-45, and 101-116.

4 For a current argument favouring a smaller assault and relatively larger suppression force, see General William E. DePuy, "One-Up and Two-Back?" Army, No 1 (January, 1980), 20-25.

5 Still today the commander of all artillery in a corps.

6 An enraged Fuller wrote: "Were these lessons and the thousand and one others which the War drilled into us at a cost of one million lives and eight thousand millions in money learned or even examined by the Higher Command? No. Ever since January, 1915, there had been preparations for demobilisation; yet no small department was created to study the War as it reeled along its sodden path so that its lessons might be valued and on its conclusion an Army might be built up which, if not war-proof, would at least be war-worthy. Thus it happened that we blindly stepped backwards and fell over the precipice of long-vanished hopes....The War Office plunged into a battle for the red coat, khaki not being considered military enough for peace-time soldiering....For once the Finance did the Army a good turn and stopped this tomfoolery...." Fuller also had some words on the army "caste system", whereby "should a gunner get what is called his 'jacket', in the military eye he is potentised, just as if he had received a graft of monkey gland. Once a black-button man, you cannot become a brass-button man, you are black-buttoned for life; and just as a mandarin's yellow button used to get him a job, so will the five black buttons on your jumper or tunic get you one". J.F.C. Fuller, The Army in My Time (London: Rich & Cowan, 1935), 37-38 and 174-175.

7 According to Major-General Sir Howard Kippenberger, a New Zealander serving in the Middle East, "There developed at one stage throughout the British Eighth Army...a most intense distrust, almost hatred...of friendly armour". Kippenberger, Infantry Brigadier (London: Oxford University Press, 1949), 180. This attitude, which was rooted in a failure in tank-infantry cooperation, contrasted sharply with the German "systematic practice of the principle of Cooperation-of-Arms" and consequent superior employment of anti-tank guns as part of the all-arms force. Major-General F.W. von Mellenthin, Panzer Battles, trans. H. Betzler (Norman: University of Oklahoma Press, 1983), 55.

8 See General Sir William Morgan, "The Revival of Battle-Drill in World War 2", Army Quarterly and Defence Journal, No 1 (October, 1973), 57-60. For renewed U.S. Army interest in battle drill see Captain John Antal, "Unit Battle Drills", Infantry Magazine, No 4 (July-August, 1983), 31-32.

9 Attack against the Reichswald by XXX (British) Corps, under command of First Canadian Army, on 8 February, 1945. It was launched after the most concentrated artillery bombardment of the war in North-West Europe yet cost First Canadian Army 15,000 casualties. Chester Wilmot, The Struggle for Europe (London: Collins, 1952), 767-769; and Shelford Bidwell and Dominick Graham, Fire-Power (London: George Allen & Urwin, 1982), 289-291.

CHAPTER SIX: TECHNOLOGY, THREAT, AND POSSIBLE NATO RESPONSE

by
Brigadier (Retd) R.E. Simpkin, OBE, MC
ELS Consultant Linguists Limited
Over Deanshaugh, Scotland

The following chapter represents the keynote address of Brigadier Simpkin to the Combat Training Centre First Annual Officer Development Study Session.

THE STATE OF THE ART

The tank is at once the dominant conventional weapons system of today's land force battle and the greatest limiting factor in that contest. Its weight, width, and complexity on the one hand dictate its segregation in terms of force structure and of engineer and logistic support; on the other they restrict every kind of mobility, not least movement speed and trafficability of the main manoeuvre force. Even the BMP-based Soviet vanguard needs the fighting power of the T72; so it must mostly go where and as fast as the T72 can go.

Thus a revolutionary change in tank design would have an extremely radical effect on the organization, versatility, and way of fighting of the combat arm. I share the view of many in the United States, some in Britain and the FRG, and most in that key country for armoured vehicle development, Sweden, who believe that such a change will occur in the Nineties. The Soviets anticipate it too but perhaps a half generation later.

Some say that the vigour with which a tradition is fought for varies inversely with its age. Certainly the tank, at a mere sixty-five, is hard to rival as a generator of irrational conservative heat. But this change <u>will</u> happen because it will be - in fact <u>is</u> being - imposed. The dominance of firepower acts on the tank and other AFVs in two ways. First and more broadly, in concert with the fixed and rotary wing air threat, it calls for more artillery and more mortars. To achieve this, combat manpower, logistic resources, and money will have to be diverted from the combat arm and its equipment. So the combat arm and <u>its</u> support will have to be streamlined. I shall work my way round to this aspect shortly.

Second and specifically, the scope and performance of anti-armour weapon systems hangs a large question mark over the survivability of the tank as we know it. In the 1980 edition of his book Tanks and Tank Forces, which the British library managed to get from Moscow for me, Chief Marshal of Tank Forces Babadzhanyan, writing as commandant of the trend-setting Malinovskii Tank Academy, speaks of the peacetime danger of sacrificing armour protection and even firepower to mobility; but he stresses that in the longer term the tank's survival will depend mainly on a "combination of mobility and concealment". Writing about the same time, Field Marshal Lord Carver, always a protagonist of protection and -- to twist the title of his book -- a "prophet of position", emphasizes the future importance of "indirect protection", by which he means not getting hit and if possible, not being seen. This is summed up graphically in both senses of the word, by Clifford Bradley, until recently chief designer at USATACOM,[1] writing in Armor Magazine. Fig. 6-1 shows how he sees the trend.

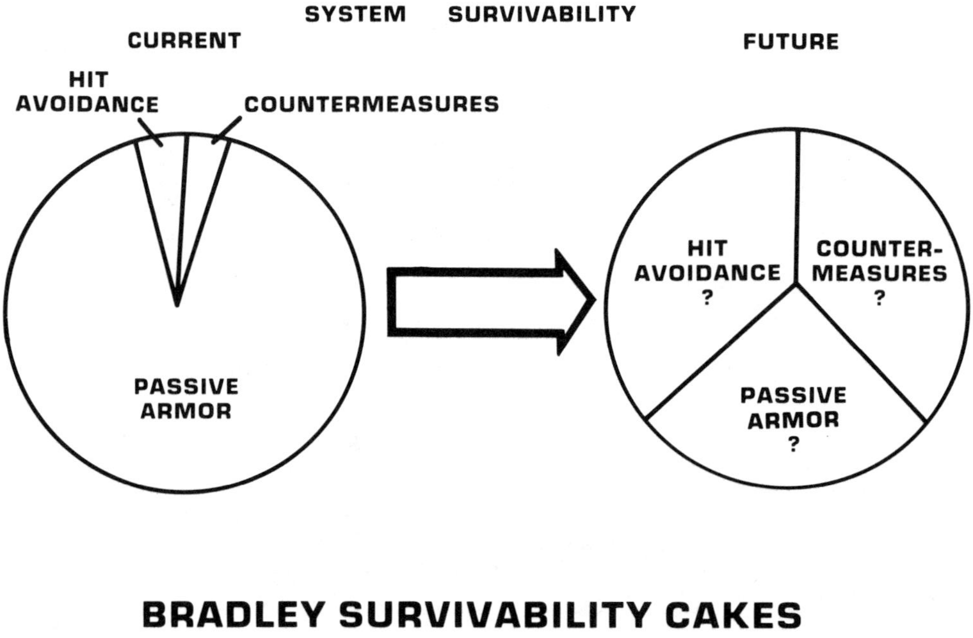

Bradley Survivability Cakes
Fig. 6-1

Following up this thought, also in Armor, I concluded that there would be a rather uncomfortable survivability gap -- the darkest slice of the cakes shown in Fig. 6-2. The left hand one is with visual target acquisition, the other with an automatic target acquisition system in the opposing tanks. When you throw in the unavoidable random chance casualties (the light grey slices), you get predicted tank losses of over forty per cent in the one case and seventy per cent in the other. Clearly we are talking not just about a difference in degree but rather a quantum jump backwards.

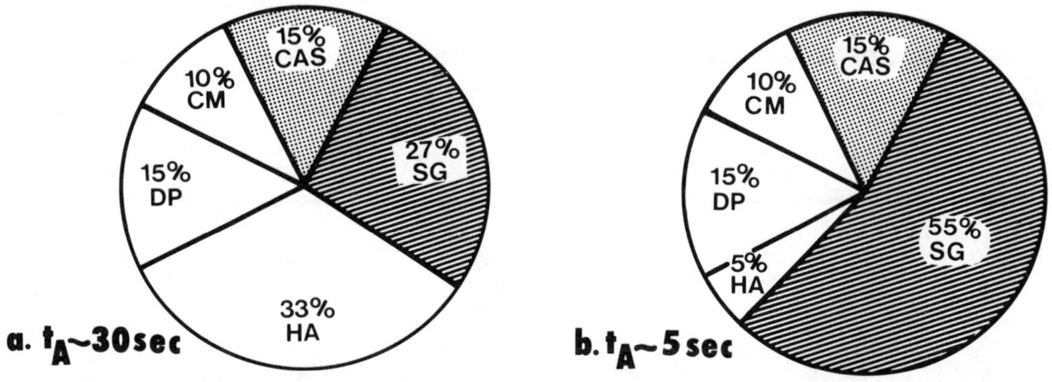

These survivability "cakes" are based on mean acquisition times of 30 seconds (a), and 5 seconds (b). The abbreviations have the following meanings: CAS, freak casualties; DP, direct protection (freak failures of attack); CM, countermeasures; HA, hit avoidance; and SG, survivability gap.

Simpkin Survivability Cakes
Fig. 6-2

The reason for it is the collapse of the <u>frontal arc concept</u>, on which the design and distribution of a tank's armour has been based ever since the first dedicated anti-tank weapons were fielded. You see the classical concept at the left in Fig. 6-3. On the right you see an estimate of the effect of the attack helicopter using nap-of-the-earth (NOE) tactics on the horizontal arc; and at the bottom a very conservative estimate of the effect of the helicopter, indirect fire systems, and downward-looking systems on the vertical arc.

Let me explain why, in fact, I stuck in that very modest figure of twenty degrees. If we think in very crude and nominal terms of thickness of vertical steel armour, the frontal arc is about twelve times as "thick" as the rest. The effect of increasing the horizontal arc by a factor of three and the vertical arc four times has thus placed the armour designer precisely back on square one!

Before I give the tank a new lease of life, let me just spell out the scope and power of this high-angle attack and the high proportion of battlefield weapons that can deliver it. Taking helicopters first, the angle of descent of a cruise-type missile fired at long range from a helicopter flying NOE will not be too great. But Gatling or chain guns of the kind mounted in the chin turrets of the Huey Cobra, AH64, and Mi24(HIND D), firing modern sub-calibre penetrators against the horizontal roof and near-horizontal glacis plate, pose a nasty problem. The Copperhead guided shell, fired from a standard 155mm gun, is a major advance, but an expensive one that depends on target marking by a forward observer.

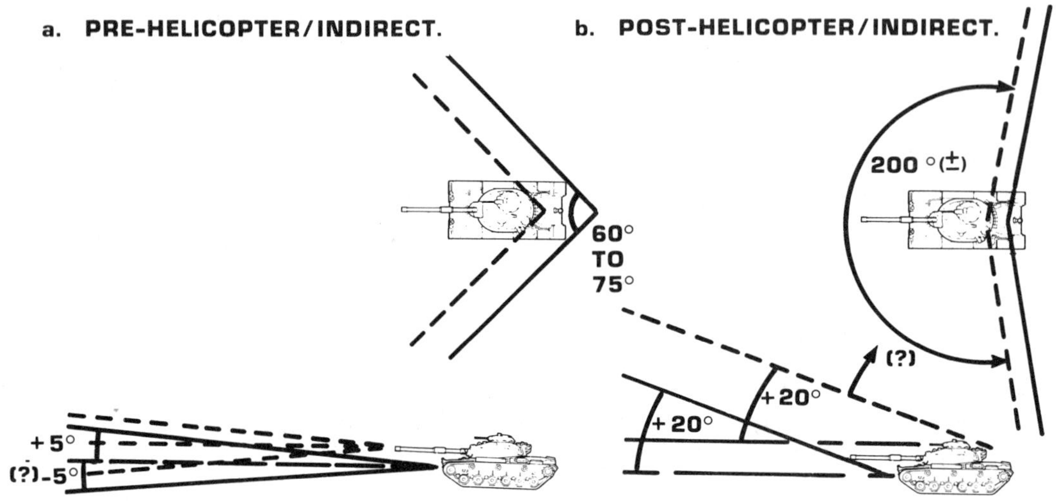

Frontal Arcs
Fig. 6-3

Of much greater interest to the smaller and more impoverished armies is the mortar bomb. Thanks to rather dramatic improvements in the potency and lethality of hollow charges, free-flight and precision-guided bombs fully capable of destroying a tank are on the market now in the 120mm calibre. It will almost certainly become possible to give 81mm or even 60mm bombs sufficient power of attack.[2] These would probably have to be freeflight, but their cheapness, the mortar's high rate of fire, and the handiness of medium and light mortars for short-range use, make this a highly attractive option.

Then there is the system (probably air-to-surface) used by the Israelis in the Lebanon, which appears to be capable of pinpoint lethal attack on the commander's cupola. I understand there are T62s and T72s with the cupola area taken out as if by a welding torch. There is already a considerable armoury of air, rocket, and tube delivered cluster munitions, carrying anything from simple two-inch high explosive anti-tank (HEAT) bomblets like the American M42, to "smart" bomblets and "mini-missiles" with dual acquisition and homing systems. By the Nineties, all these sub-munitions will be capable of delivering self-forging fragment (SFF) attack as an alternative to hollow charge.

Before I go on to discuss the counters to these developments, let me stress the common technological base of all tank-developing countries. Bright-eyed and busy-tailed with expectation, and armed with a clipboard recharged with blank paper, I sat down one day last summer to devil the technical section of the book by Babadzhanyan I mentioned earlier. I finished it a day or two later with precisely three lines of notes -- of which more in a moment. Give or take a few bits of fancy mathematics, the rest could have come straight from the teaching material used at Fort Knox, Bovington Camp, Munsterlager, Saumur, or Skoevde. Whether we like it or not, modern AFV technology is firmly rooted in the foundations laid by the Third Reich. As I see it, the Soviets currently have an edge in gun-ammunition systems and probably in running gear; but are behind the FRG in power trains and fire control, and behind the United States in optronics, particularly night vision.

With this in mind, let us look at the tank designer's response, which evidently needs to be as dramatic as the enhancement of the threat. On-vehicle countermeasures may be a technical possibility. But quite apart from the design and cost penalties, they have the grave drawback of disclosing their user's tactical layout. And any aggressive system presents a hazard to friendly forces. So the designer has little choice but to put most of his money on hit avoidance, and then to see what he can build back in in the way of direct protection -- armour, that is. Even the dinosaurs of the Eighties, like Leopard 2 and its look-a-like M1 Abrams, are designed around agility -- the reduction of firing and movement exposure times. But the vast targets their turret fronts and sides present even when they are in a fire position, and the vulnerability of their roofs, lay their survivability in the face of the Nineties attack open to question. Many think, though I do not, that the Soviets have gone, or are going, the same way on the tank we call "T80", with a spaced-armour turret on the T64 hull and a weight of forty-eight tonnes or more.

The fact is, the swing of the tank/anti-tank pendulum still prevails. You will all have heard of fibre optics from the technique's medical applications. Those designing armoured vehicles for the Nineties will have at their disposal a development known as the "optical slip-ring", which allows an optical image to be passed from a sight head on top of a traversing turret or mounting to a fixed eyepiece in the hull on which it rotates. There are already a range of optronic systems that allow the designers to do this trick electronically. At the moment these optronic images are inferior to optical images in one key respect. Optical systems have higher resolution and are thus better for visual target acquisition in the condition represented by daylight obscured by about one half cloud cover and, literally, the fog of war. But for the Nineties optronic systems, combining inputs from a number of different sensors via an image processor, will almost certainly equal or surpass optical systems in this respect too. Either or both of these options will allow the designer to put the whole crew in the hull.

The vehicle in Fig. 6-4 is a Swedish test rig that mounts the NATO 105mm tank gun on the hull of the German Marder infantry fighting vehicle. This is an Eighties design with a direct optical system, so the gunner is located in a one-man turret under the gun. Despite this, the dramatic reduction in the volume of the armoured envelope stands out a mile. More striking still is the virtual non-target offered when the vehicle is in a fire position (Fig. 6-5). This historic test rig has now been dismantled -- as has its German counterpart.

Marder 105mm Test Vehicle
Fig. 6-4

Marder 105mm in Fire Position
Fig. 6-5

Sweden's next move, again an Eighties project known as UDES XX 20, was to combine the above layout with "double articulation" (in which the driver has control of both horizontal and vertical angles between the two parts) in an anti-tank and support gun for its Norrland infantry brigades (something that could interest Canadians). The gun is the German 120mm smoothbore of Leopard 2, with a special Bofors muzzle brake, and an autoloader fed from external magazines on the rear of the hull. The train weight is about twenty-three tonnes (Fig. 6-6).

Swedish UDES XX 20
Fig. 6-6

Still with Sweden, shown in Fig. 6-7 are three concept studies for the Nineties main battle tank successor, known as UDES 40. These are all crew-in-hull concepts with the 120mm gun in an external mounting. One has a yoke mounting; one a rotating pillar; and the third a pillar which both rotates and retracts the gun to hull-top level.

Very similar both to the Swedish concepts and to some of Cliff Bradley's sketches is the outline concept shown in Fig. 6-8, which Armor Magazine artist Mark Irwin worked up from some roughs of mine, and upon which I should like to expand. Of note are the multisensor heads which move independently of the gun (here a 105mm); the Swedish type magazines and autoloader; the crew of three in line, with the driver on the left and identical commander's and deputy commander's stations centre and right; also the flaps covering the grouped fillers. If this was built as a thirty ton-ish vehicle, I would envisage it with Leopard 2's power train, giving it a power:weight ratio of some fifty brake horsepower per tonne, which is formidable.

Swedish UDES 40
Fig. 6-7

This approach tackles hit avoidance first and foremost by cutting the whole armoured volume right down and particularly the exposed target area presented in an observation or fire position; it also reduces the chance of detection and the target size when the vehicle is moving behind a skyline or through low cover like scrub. In a position of observation only the sensor heads are exposed. Really high automotive performance, with ride to match, additionally minimises firing and movement exposure times.

To return to direct protection, one can see by looking at the main pod of this vehicle in plan view and section (Fig. 6-9), how an integrated compound armour system is used to optimise direct protection. Behind steel plates and a compound array, the fuel tanks (F) and the power train are used to help protect the crew. Designers of future AFVs will have to abandon the ship-like approach of an armoured envelope and adopt the aircraft principle of protecting vital areas against specific risks.

The exploded view in Fig. 6-10 indicates how a modular design of this kind could provide the basis for an AFV family. With an appropriate interface panel (number 4) you can link the crew-automotive pod (3) to functional modules (5) in weapon platform or "tin box" configurations.

To recap briefly, the change in the anti-armour threat means that future AFVs will rely for survival mainly on indirect protection -- not getting seen and not getting hit. This means that tanks have got

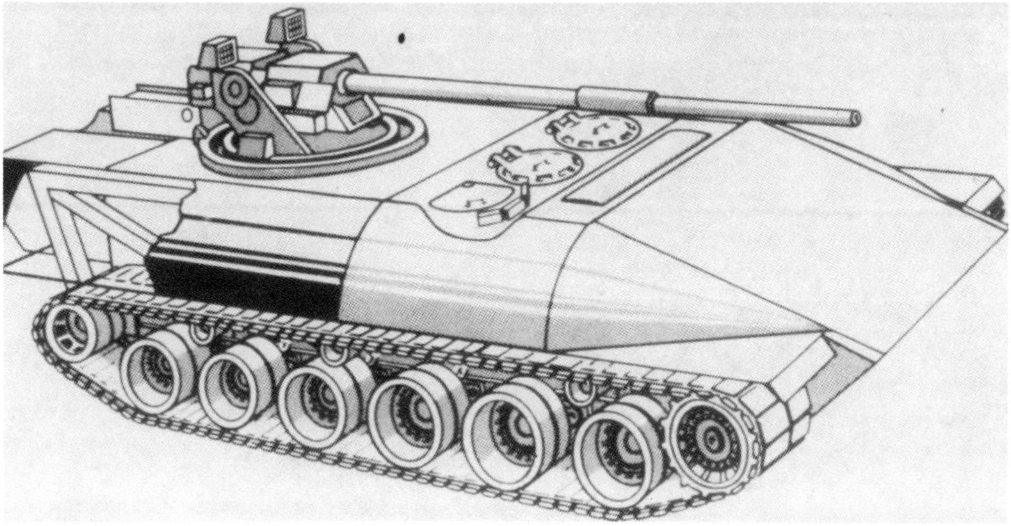

Simpkin IFV Concept
Fig. 6-8

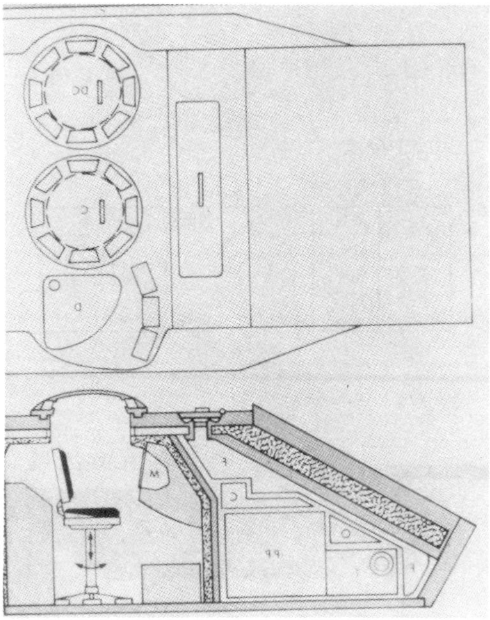

Plan and Section Views
Fig. 6-9

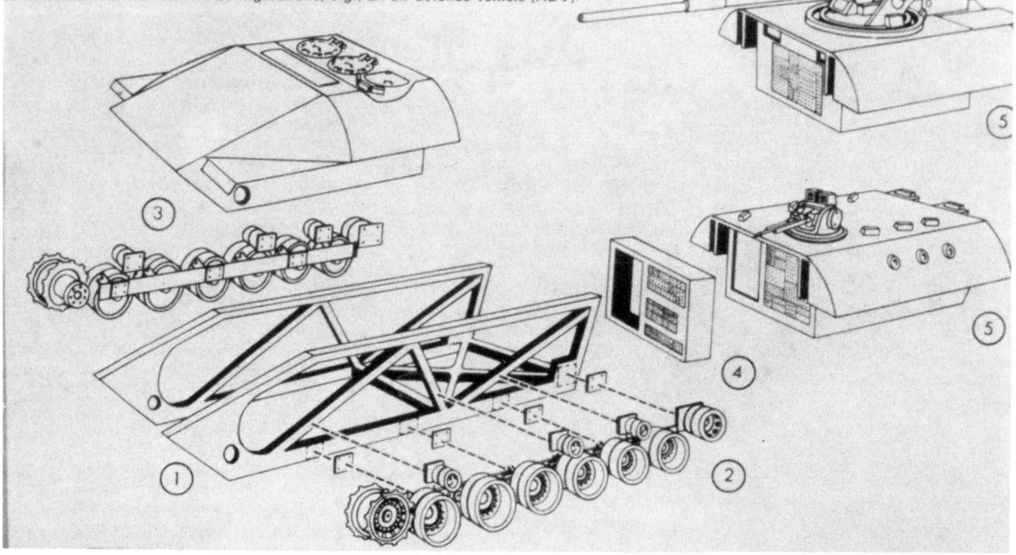

The pod concept. The constant elements consist of a "chassis" frame (1), running gear (2), and a heavily protected crew/automotive pod (3). To this are added an interface panel (4), and one of a number of functional pods (5). The functional pods may be based on an external mounting (tank destroyer or fire support tanks), a compartment with or without a slaved weapon pod (IFV, CSV), or some combination of the two configurations; e.g., an air defense vehicle (ADV).

Pod Concept Exploded View
Fig. 6-10

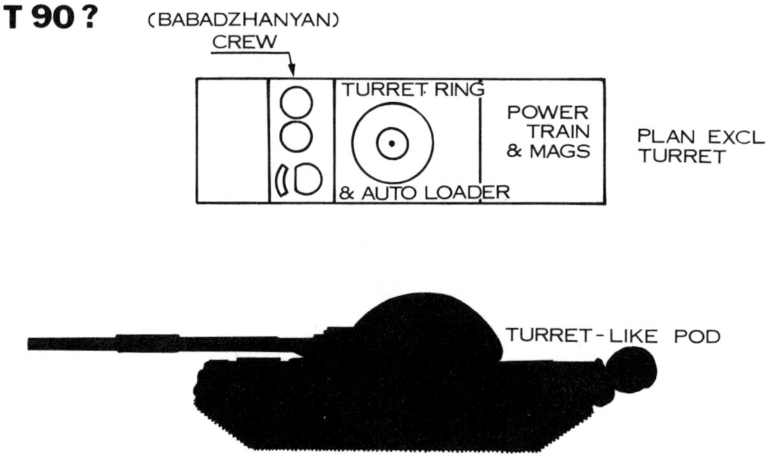

Soviet T-90
Fig. 6-11

to be much smaller, lighter, and handier. The technology to make this change without sacrificing any key characteristics is available, or just about so. Once this is done, it becomes cost-effective to incorporate the main battle tank's successor into a family of AFVs.

In case you should think that all this is personal pie-in-the-sky or some dreamy slice of Swedish caramel, I would refer you to another option shown in Fig. 6-11. This was the subject of the one research photocopy and three lines of notes that, as I mentioned earlier, I took from the technical section of Babadzhanyan's book. The reasons for the residual turret-like pod instead of an external mounting are probably the Soviet philosophy on ammunition protection -- an interesting variation on Russian roulette -- and their existing design of autoloader, which works well on the T72 though it is known to have eaten at least four gunners on the T64. This could be the germ of "T90".

U. S. Light Tank
Fig. 6-12

Finally, then, what weight will be the trend of the future? In the case of "T90", which I reckon would use a development from the T64 hull, I estimate something under forty tonnes; like the Swedish "S" tank, it will probably swim with a screen. For UDES 40, the Swedish general staff were offered thirty or forty tonne versions; they opted for forty because of the much higher level of direct protection. If my analysis is right, one would also expect to see the Soviets field a mobile, protected high-performance gun based on the BMP. At the other end of a more or less continuous spectrum, you have the United States test rig, HSTV-L, now evolved into a light tank of just under fifteen tonnes with the ARES "Super 75" burst-fire gun (Fig. 6-12). At one of these extremes, or somewhere between them, lies the survivability pay-off point, with the best combination of direct and indirect protection. Somewhere in there too lies an overall optimum.

THE SOVIET LAND THREAT

Not long ago I read a report in Management Today, the British Institute of Management's journal, on a lecture by a distinguished retired officer. "Like all these military men," it read, "he started with the threat." I read on to discover why this was such a bad thing to do and realized that the reporter was identifying the threat with the competition. In fact, of course, it's the market. So let's crack on with a bit of market research.

I have spent a good deal of time in the last fifteen months doing research and the impressions I've ended up with are as different from the standard NATO intelligence picture as chalk from cheese. As most of the point about the conventional land threat is its scale, however, it is rather difficult to relate it to the theme of this symposium, small forces and low level tactics. What I shall attempt to do, therefore, is to offer you a thumbnail sketch of the main features of the threat as I now see them. Next, touch on the low-level tactics of first contact and initial follow-up, and then try to highlight the opposition's weaknesses. From there I shall proceed to consider how best to exploit these weaknesses.

The one new element in the Soviet order of battle (ORBAT) is the airborne assault brigade (Fig. 6-13). This is entirely manned by members of the airborne forces, who come third in the pecking order for manpower selection as opposed to the land forces' sixth, and who sponsor the increasingly important special forces. When fully deployed, these brigades will be allocated on a scale of one per front (rough equivalent of a NATO army group) and one per tank army formation (equivalent of a NATO corps) which will place ten in the forces facing the NATO centre. Soviet thinking, based on the movement of masses, regards mobility as a concrete element in combat worth. From this principle, the nomenclature, and other evidence, they seem to regard these helicopter brigades as the air-mechanized equivalent of a tank division. The Soviet operational manoeuvre group (OMG), their latest version of the spearhead, is normally formed by a reinforced tank division. It thus seems clear that an airborne assault brigade could itself form an OMG, very possibly with a tank division as follow-up; or it could be used for the immediate and potent reinforcement of a tank division in the OMG role. These brigades are most unlikely to be committed on tactical missions. The helicopter assault regiments of tank divisions, or front rotary-wing resources, in either case lifting specially trained motor rifle battalions, are available for tactical missions.

So once a breakout from operational depth is established, what you are going to meet at the head of each axis is either a tank division, or an airborne assault brigade, or both. Let us now examine a little more broadly the paths these spearheads might trace.

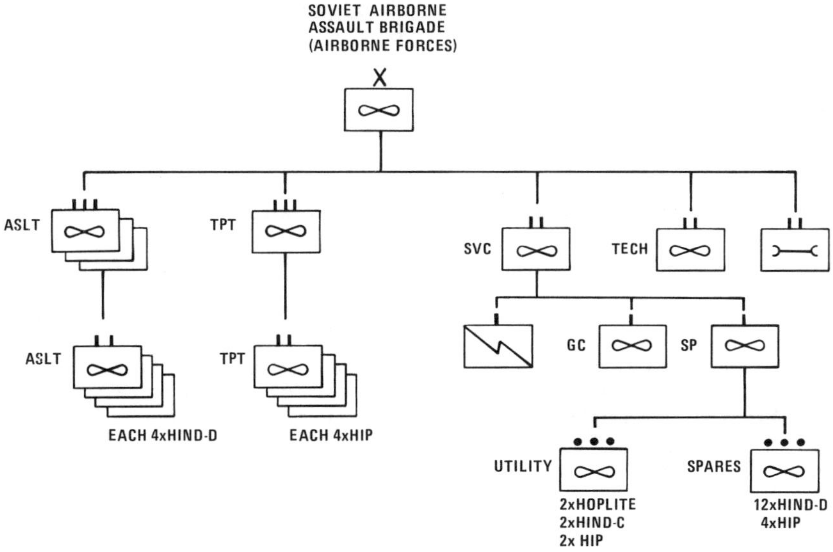

TOTAL HELS (EXCL HQ): 60xHIND-D, 20xHIP, 2xHIND-C, 2xHOPLITE (84)

Soviet Airborne Assault Brigade
Fig. 6-13

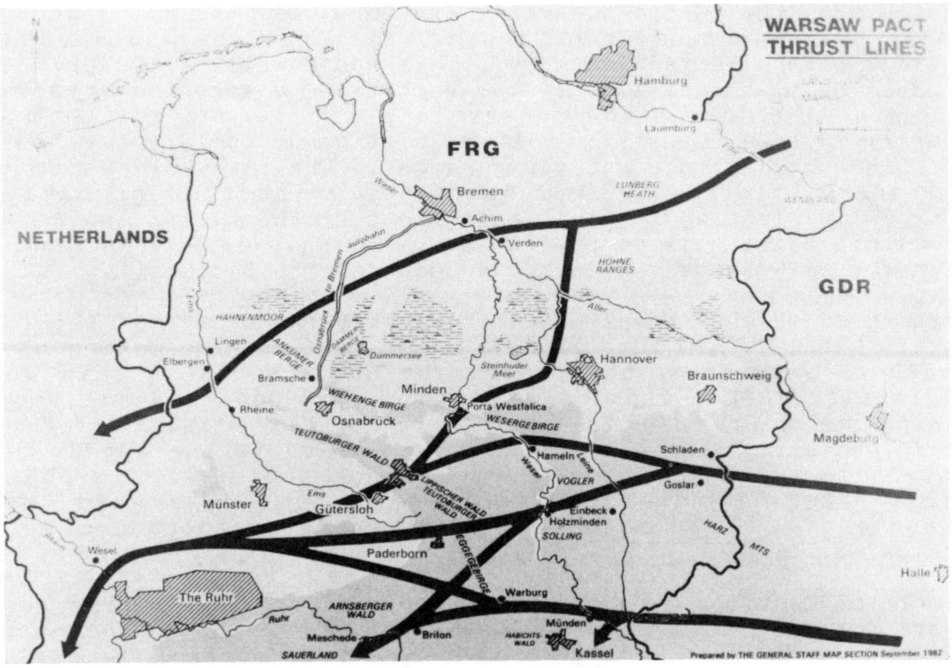

Warsaw Pact Thrust Lines
Fig. 6-14

The schematic in Fig. 6-14, which comes from a recent article of mine in the British Army Review,[3] indicates the main Warsaw Pact thrust lines. It does not include either amphibious landings on the North Sea coast or airborne divisional operations (with fixed-wing transports) that would almost certainly go in behind initial strategic objectives, such as Bonn or Brussels, to isolate and sandwich them.

Quite apart from the intelligence leak, deliberate or otherwise, of a few years back, there are many reasons to suppose that, if the Warsaw Pact launches an offensive after a period of tension and build-up, their main effort will come down the northernmost of these lines along the historic east-west invasion corridor. Astonishing as it may be, there seems to be genuine doubt whether their middle-piece officers, perhaps even up to regimental commanders, are capable of dealing with the intricacies of the more southerly terrain, although they can take water obstacles like the Weser in their stride. Then again, the North German Plain is well-suited to the Pact's drill-book way of doing things. The communications and terrain on both sides of the Inner German Frontier (IGF) are nicely matched to their echeloning and formation relief procedures. And the Pact may well feel increasingly that a sea flank with the possibility of landings is an advantage. As my arrows suggest, there is no great problem about swinging down by Elbergen, where the River Ems and the Dortmund-Ems Canal abut one another, onto the Rhine at Wesel. This gives access behind the Ruhr to Bonn.

On the other hand the Kassel Gap, the southernmost of these lines, clears the Elbe within their own territory; presents the shortest distance from the IGF to the Rhine and Bonn; and offers a good deal of what the Americans, the Germans, or ourselves would consider excellent offensive tank country. This could well be the principal, or indeed only, thrust line if the Pact tried a surprise attack with their standing forces moving straight off manoeuvres or out of barracks. Such a thrust would be made at great speed, very possibly over a chemical, if not a nuclear carpet. The aim would be to reach the Rhine within twenty-four hours, by which time probably two airborne divisions would have landed west of the river; and to seize Bonn within thirty-six hours. To my mind this is a very dangerous option for NATO and a very tempting one for the Soviets.[4]

One thing of which I think we can be fairly sure is that the "heavy break-in battle" NATO intelligence spent so much time making us love to hate is a thing of the past. The Soviets reckon this was an expedient forced on them by limitations in equipment quality and training standards. The "revolution in military affairs" of the late Fifties and early Sixties shifted the Soviets to the heyday of the battlefield nuclear weapon. And the non-nuclear first phase of Exercise "Dnieper" in 1967 was the point of departure for the non-nuclear all-arms concept that is now just about evolved and promulgated.

One word of caution, though. The rather deliberate all-arms concept appears dominant, an indication ostensibly borne out by the increase in the proportions of artillery and infantry in the tank division. For the technological reasons I opened with, this latter trend makes absolute military sense. The studies I have made, moreover, leave me the reasonably firm impression that the all-arms armies and their operations are, in effect, a vast mobile launching pad for the raid forces and the tank or helicopter-based operational manoeuvre groups.

This emphasis on the spearhead leads me to the last general point I want to make before panning in on some tactical vignettes. The principle of exploitation and reinforcement of success is an absolute. Totally unlike anyone in the West except perhaps those other apostles of mobility, the Germans, the Soviets are prepared to funnel down to a company attack frontage with wide open flanks if only one leading company is still moving forward. The deduction I want to make from this is that a positional defender has got to win the first and possibly the second encounter. If he can do this, the defender can at least expect the opposition to go the other way. And that seems to me a good practical aim for David trying to ward off Goliath. It multiplies up in just the way the Soviets like least and is, I believe, the sort of link between the massive spearhead of the offensive and the small unit game with which we are concerned.

If we now turn to detail, a favourite Soviet tactic is the encounter battle or, if you prefer, a meeting engagement. They would dearly love to catch NATO on the way from barracks to its deployment positions.

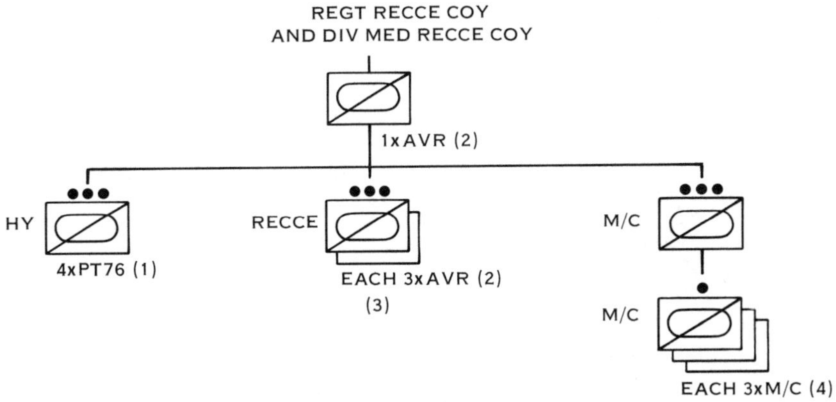

NOTES

1. MAYBE NOW T-62A PENDING NEW "LT TANK"
2. WAS BRDM 2 OR 3, NOW BMP "KORSHUN"
3. OR MAYBE 3 PTLS OF 2
4. OR MAYBE 4 PTLS OF 2

Regimental Reconnaissance Company/Divisional
Medium Reconnaissance Company
Fig. 6-15

First among things you may meet is the regimental reconnaissance company or divisional medium reconnaissance company (Fig. 6-15). An interesting sidelight is that reconnaissance units are directly controlled by the "second staff officer", that is the senior intelligence officer or chief G2, of their headquarters. Patrols from these units may be fifteen to twenty, or even fifty kilometres ahead of the vanguard or line of contact. Their orders are normally to attack only nuclear delivery means, identified key headquarters -- and, rather interestingly, delivery systems for scatterable minelets.

I said earlier that an operational manoeuvre group might be formed by a reinforced tank division, and you may have wondered "why reinforced?". One or more raid forces may move with an OMG and peel off to deal with specific targets en route or to a flank. Or they may sally forth independently. And for the future one can see some extremely interesting interaction in depth between raid forces and airborne assault brigades. Raid forces, normally ad hoc, are apt to consist of: a BMP-mounted motor rifle battalion with a tank company (and here one sees the screaming need for a powerful light tank); reconnaissance and NBC surveillance sub-units; an SP artillery battalion, probably with 122mms or multi-barrelled rocket launchers; an air defence platoon or company, certainly with ZSU24s and probably with Gaskin or its successor; extra forward observers to handle heavy tube and rocket artillery; and a platoon of sappers with AVLB (armoured vehicle launched bridge) and/or tracked SP ferry/bridge equipment.

A TYPICAL SOVIET
RAID FORCE

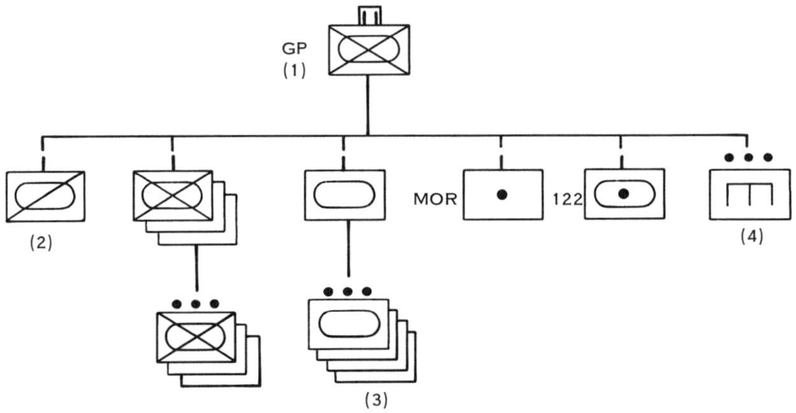

NOTES

1. BMP
2. REGT'L COY OR DETS
3. WAS 3 PLS
4. PROBABLY PLUS AEV (ROUTE CLEARANCE) AND/OR AVLB AND/OR SEC SP TRACKED FERRIES.

Typical Soviet Raid Force
Fig. 6-16

Except that the battalion may be mounted in wheeled BTR60 or 70 series APCs, this is very much the same structure as an all-arms vanguard; of course, the vanguard has the rest of a motor rifle regiment group to back it up. In either case, the order of march of the <u>point company group</u> will be similar to that shown in Fig. 6-17. It will have the reconnaissance patrol a bound ahead, and a close reconnaissance patrol, often one tank, at its head. The point does not employ fire and movement, but motors continuously in column on its route.

The advance guard of a <u>tank</u> division will be a tank regiment, with a tank battalion group as vanguard. The vanguard's artillery battery, in direct support and under command for movement, is likely to be a multi-barrelled rocket launcher (MBRL) battery as opposed to 122mm SP guns. The brief but intense salvo of the MBRL is considered to be better suited to the tempo and shock effect of tank forces. Otherwise the point is very much the converse of the all-arms point, and we need only look at the leading elements.

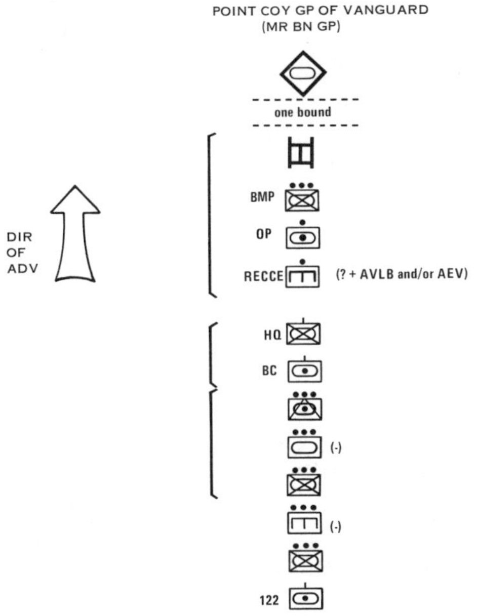

Point Company Group
Fig. 6-17

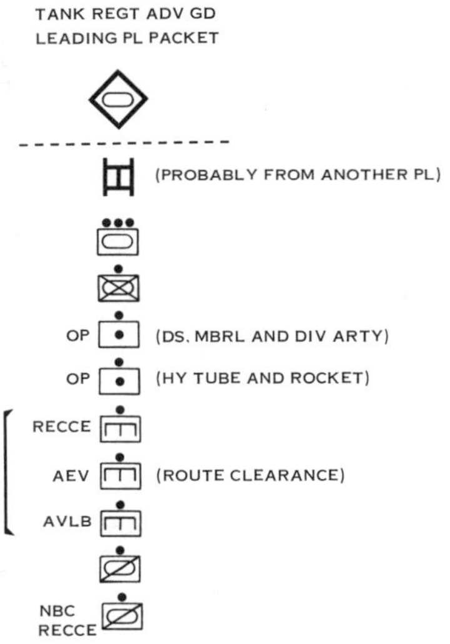

Lead Tank Platoon Packet
Fig. 6-18

In the leading tank platoon packet (Fig. 6-18), there will probably be an NBC surveillance patrol and possibly a "reserve" reconnaissance patrol. The sapper reconnaissance detachment will be boosted by an armoured engineer vehicle (AEV) set up for route clearance and probably by an AVLB. When you think that this grouping has only a motor rifle section to protect it, you can see one good reason why the Soviets are following the example of history and increasing the proportion of infantry in their tank formations.

To return to the all-arms point company. On contact this will deploy into column of platoons (or platoon groups), and pound the flank of the opposition, with the infantry fighting mounted, until the rear platoon is just short of the front of the position. By then artillery covering fire will have opened, and the company attempts an enveloping attack. The idea of turning is absolutely fundamental to

Soviet tactical thinking; in other words, they really do go right around the back (see Fig. 6-19).

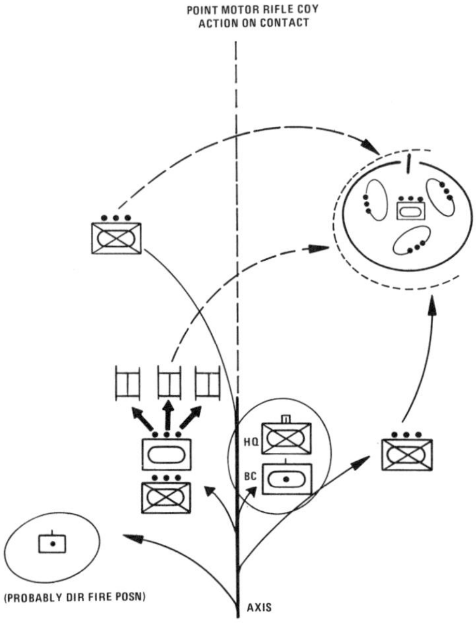

Basic Soviet Encounter Tactics
Fig. 6-19

If this fails, or if the defensive position is patently too large, the vanguard commander may commit the rest of his battalion group. More probably, though, the advance guard commander will mount an attack with another battalion, very possibly ordering the vanguard and his third battalion group to continue the advance.

Leading Soviet elements tend to go on motoring under controlled movement conditions in battalion or even in regimental column, until they expect or make contact; or deploy for some planned purpose such as crossing an obstacle. Deployment is then ordered in stages -- say, from battalion column to column of companies, column of platoons, line of platoons and finally, especially for tanks, line. By the same token they get motoring again and progessively reform column once they have cleared through their final objective. It is a totally reversible procedure.

Thus both the advance and the approach march are executed in column. So they do, in fact, deploy for an attack "straight off the line of march". Now I have tried that and I know of Americans and Bundeswehr people who have tried it with their battle groups and combat teams and invariably the result is totally delirious. One is prone to ask why the Soviets are cleverer at it than we? What is less obvious until one goes into it, is that before this impressive display they have gone through a complete sequence of battle drills -- movement planning, regrouping and movement orders, "R" (reconnaissance) groups, "O" (orders) groups, etc.

Four or five weeks ago I was studying a series of articles in Voennyi Vestnik (their Military Herald, a monthly professional journal aimed at junior and middle-piece officers, and perhaps senior NCOs). These articles, from the "all-arms battle" series, were mostly headed something like "the speedy attack", and most singled out a battalion or a regimental headquarters for praise. I was flabbergasted to find times from receipt of mission to H-hour of between sixteen and twenty-two hours for a straightforward battalion attack. At that speed anybody can be dead clever and get their beauty sleep into the bargain. I have not yet unravelled this mystery, but, while I am the last person to decry the opposition, my research has led me to suspect that all the Soviet emphasis on speed and tempo is a shade suspect by German or Israeli standards.

A battalion will probably attack with two rifle companies and the tank company less a platoon in the first echelon; the third company and a tank platoon in the second echelon or tactical reserve; and the statutory but small anti-tank reserve. An echelon is a depth force pre-tasked for a mission, for instance a follow-through objective, and thus not fully available to the commander as a reserve. The tendency now, particularly at lower levels, is towards a single echelon, with a good deal in the shop window and the kind of tactical reserve one might see in the West.

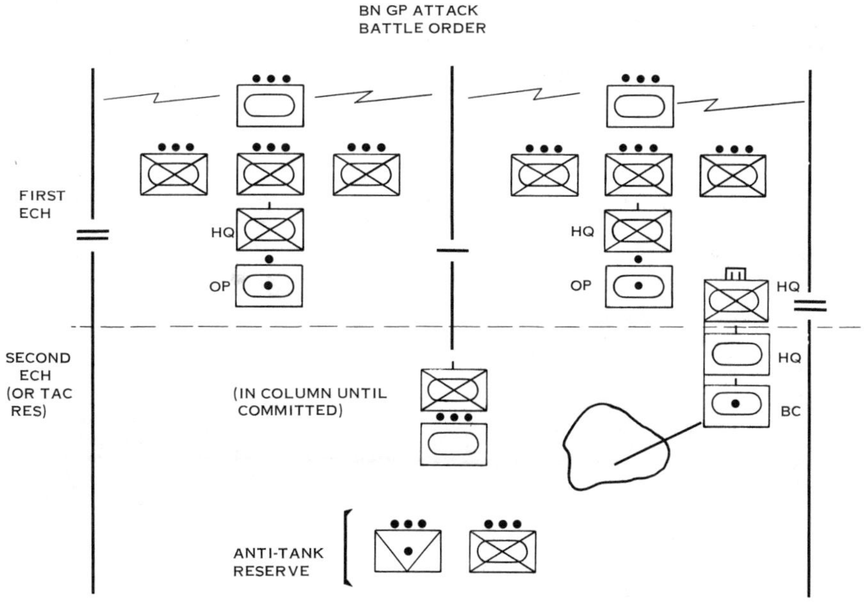

Soviet Battalion Group Attack Formation
Fig. 6-20

Depending on the ground and the weight of artillery defensive fire, leading BMP companies will probably dismount about two hundred metres behind the leading tanks and perhaps four hundred metres from the initial objective. Wheeled APC battalions debus further back and attack with the tanks supporting the leading infantry (the APCs going temporarily into a rear rally). The subject of where to dismount remains a topic of hot controversy amongst the Soviets.

The artillery covering fire plan is meticulously timed. For a battalion attack, it usually starts between five and twenty minutes before H- hour. The battalion will probably have its regiment's organic artillery battalion in direct support, with certainly the "divisional artillery group" and probably an "army artillery group" in support. Those of us who regard artillery fire on dug-in troops purely as an instrument of neutralization need to bear in mind that, based on the historical excellence of their artillery and Second World War data, the Soviets regard it as a destructive arm and talk of "artillery offensives". The norm for, excluding smoke, a battalion attack against a hastily prepared defence is about fourteen hundred rounds of 80mm calibre and upwards per hectare (2.5 acres) of occupied ground. This is a density of one projectile per seven square metres; and two or three norms may be ordered if the situation and the importance of the objective are considered to warrant it.

Quantities of smoke in terms of ammunition scales don't mean too much, at least to me. But it is normal to lay a lateral smoke screen (partly from vehicle exhausts and "large smoke canisters" seen carried on some tank and other AFV turrets) in front of the start line; to rely on masking screens rather than natural cover to aid fire and movement, particularly debussing; and either to lay blinding screens, or to mix smoke in with the HE, to cover the attack over the last two or three hundred metres.

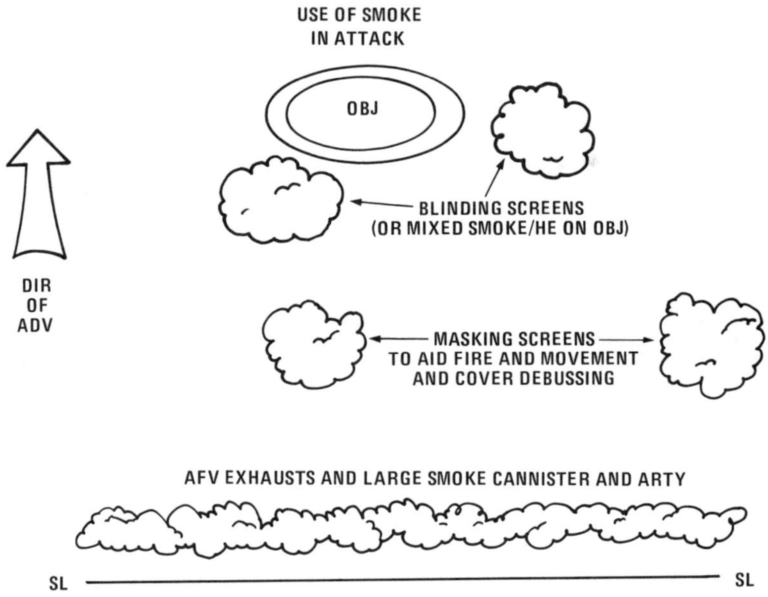

Soviet Employment of Smoke
Fig. 6-21

Finally, I want to take a quick look at river crossings. The Soviets distinguish three types of crossing operations. One is the bounce crossing, the seizure of an existing crossing by a heliborne coup-de-main party, a reconnaissance unit, or the point company -- perhaps after having swum across the river at another point in its vehicles. Then there is the immediate, or as we might say the "hasty" crossing; third, of course, the deliberate crossing.

Before we explore the immediate crossing, let's just review the way Soviet commanders plan a large-scale crossing. This gives a very good insight into the way they think. They do not see the ground unrolling before them and their tanks rolling over it. They take a bird's-eye view and, by a kind of mental time-lapse photography, envisage the mission as accomplished. They seem to work outwards from a given line, the line of contact or in this instance the water obstacle, in a kind of mirror-image way. Depending on the level, the maximum length of the "lever arm" from this line may be anything from fifteen to three hundred kilometres.

The Soviet commanders will first define the ultimate higher tactical or operational objective, which is likely to be "balanced", so to speak, by the concentration area. They will then define initial and secondary tactical objectives, whose depths roughly "balance" the distances to the rear of the waiting area (or assembly area) and the attack position (FUP) -- or in the case of a crossing, the preparation area. The aim as they see it is not to seize a bridgehead, but to get a force rolling again on the other side of the obstacle.

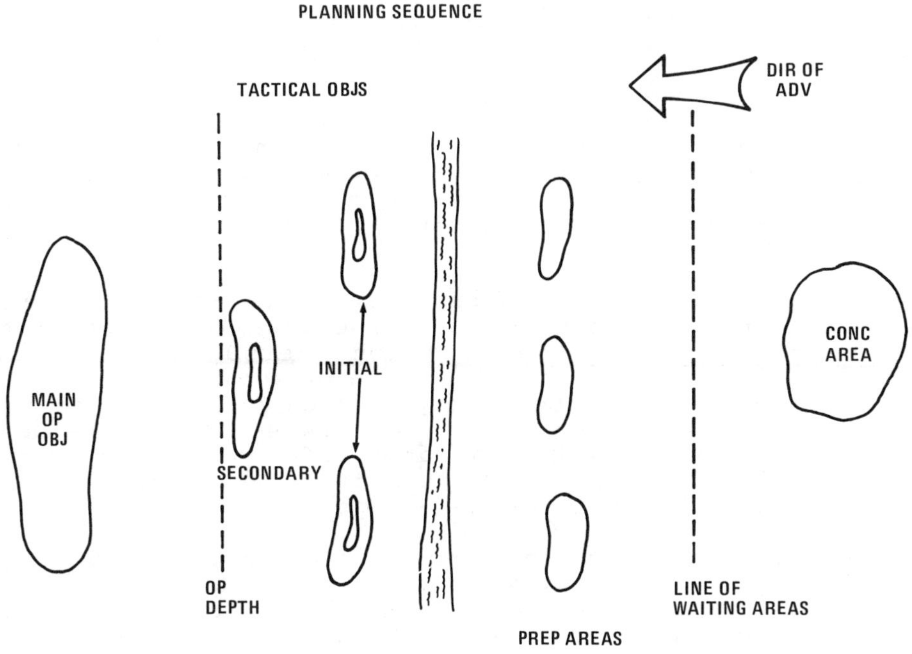

Soviet River Crossing Planning Sequence
Fig. 6-22

The immediate crossing would probably be a regimental operation executed by a battalion group -- possibly the advance guard but more probably the second battalion in the order of march. Suitable engineer resources, almost certainly tracked SP ferries, move at the rear of the advance guard or well up in the next major serial. The Soviets appear to be very much better than any Western army in ensuring they have suitably equipped sappers well forward.

Let us now suppose that a vanguard is deployed on the near bank, probably with its tanks covering across the river and its infantry, with patrols already across, perhaps maintaining a series of probing crossings in their BMPs. The crossing battalion will establish a combined tactical headquarters. Direct support artillery will be sited well forward (preferably close to this headquarters, so that it can reach out) and doubled up, so that support can be maintained while one artillery battalion is crossing. Similarly, the forward army artillery group will probably close up onto the divisional artillery group. Fire planning is based on battery targets, with about fifty per cent superimposition. Following this pattern through, the crossing battalion's tank company will probably line up with the vanguard's tanks, covering across the obstacle, until it can start crossing; alternatively, if it is going to have to snorkel, it will be held in a preparation area.

The first echelon will consist of two companies, with sapper reconnaissance and working parties. They will probably cross two platoons up, with the third poised to reinforce success; but will go three up if there are enough potential sites. Their initial mission is to secure the far side of crossing sites so that tanks can start crossing. Their secondary objective will be in some depth, and their mission at that stage will be to resume the advance.

The second echelon will be poised in a preparation area or, if possible, in dead ground forward of it. It will include the sapper ferry unit and heavy equipment. The tank company will move with it, or, if already deployed, join it as it moves forward. This echelon's primary mission will be to reinforce an infantry crossing and establish at least one tank crossing. This done, it will press forward and resume the advance.

The key point is that both the crossing battalion group and the original vanguard, probably in that order, will be unequivocally tasked to press on. If there is considered to be a risk of counter-attack, the earliest possible time of this is calculated. If time allows, the third battalion group will be moved in to secure the bridgehead until it can be relieved. Otherwise, or if the advance guard is from a tank formation, a motor rifle battalion, probably from another division, will be lifted in by helicopter using the front's (or tank army's) tactical helicopter lift.

I think you will agree that the principle of reinforcing success even -- as we should see it -- at very high tactical risk is going to crack any positional defence sooner rather than later. Some form of aggressive defence is evidently called for. I hope I have said enough to highlight my impression that persistent harassment is likely to disrupt the functioning of the Soviet machine. For instance, as I mentioned just now, the Soviets are absolutely

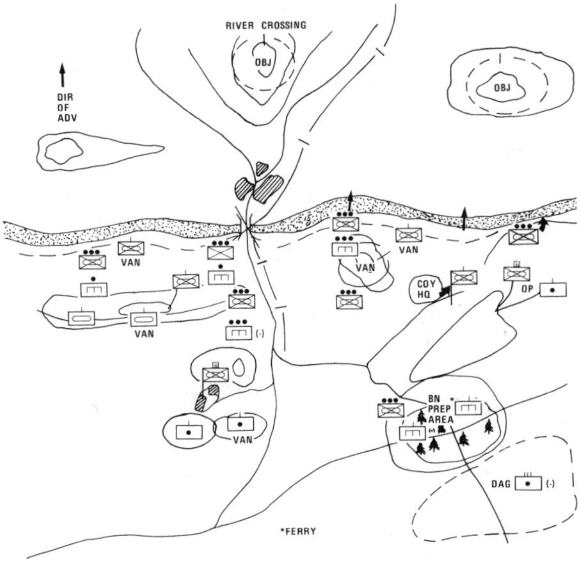

Soviet River Crossing Deployment
Fig. 6-23

paranoid about scatterable minelets. These pinpricks should aim at imposing stress on battalion headquarters, whose staff are minimal in quantity and, with the possible exception of the commanding officer, apt to be limited in quality.

By the same token, I hope I have brought out that, despite all the noise about speed and flexibility and the emphasis on the all-arms battle, the all-arms formations are really quite ponderous and rigid, while for the tank and helicopter units the name of the game is high tempo. It thus makes extremely good sense to find some way of progressively deepening the defence, so that the Soviets never achieve "operational depth" and let their tank force loose.

A POSSIBLE RESPONSE

Before trying to answer the question I have posed, I would ask you to tuck two general points in the back of your minds. The first is that the medium-sized members of NATO are just going to have to devise armed forces suitable both for the battle of Europe and for intervention outside it, to safeguard vital Alliance and national interests or to support the UN. Only the superpowers can any longer afford more than one set of soldiers or the weapons and equipment for such soldiers.

My second point has to do with structure. Within a conventional division we normally find three combat arm brigades of any kind you like, a brigade's worth of artillery, and a battalion or so each of reconnaissance and sappers. If we now increase the artillery to nine or ten major units, which we need to do, we have a problem. If we then replace the armoured reconnaissance battalion with a helicopter brigade, we are getting out of scale. If the three basic brigades need to be of two or three different kinds, we're in worse trouble still. Yet if we extract any one of the elements I've mentioned, we've lost the integrated structure needed for the "club sandwich" land force battle. From discussion with a number of people, however, it would appear that any permanent divisional structure becomes either unbalanced or unwieldy. The long-term answer may be to make the brigade the basic functional formation, and group brigades ad hoc under two-star task force headquarters.

To return to the question at hand, I would like to refer to another concept which, rather surprisingly, the Soviets write about more coherently and rationally than anyone else -- the diminishing "usability" of organized armed forces. War between organized forces, they argue, will cease to be an active instrument of policy. On this line of thought, if it decides to attack in Europe, the Pact needs to achieve its aim before NATO can mobilize or even warm up its decision-making machinery. The other side of the coin, however (at least to my mind), is that containing the situation and gaining time becomes a proper aim, even _the_ proper aim, of the defence. A more or less linear _positional defence_ will not do, because the Pact will soon break through it somewhere and run wild. Then Europe is headed straight for the nuclear threshold.

A mobile air-mechanized defence, of the kind I developed in my book _Antitank_, will not do for three reasons: few countries can afford it; it gets out of scale for small armies; and above all, it resonates with the offensive to expand the dimensions of the battlefield and accelerate the tempo of the battle -- this in an era when the depth between the IGF and the Rhine is becoming of only tactical significance. Another and rather similar aggressive defence, General Starry's "airland battle", is open to the same criticism. On top of this, it seems dependent on weather and the air situation; and its positional elements appear deployed to create air targets rather than to achieve their utmost in the land battle itself.

So we come to the _operational counter-stroke_ to the Oder or wherever, favoured by some _Bundeswehr_ tank men and an increasing number of British officers. Here again the objections are threefold. If NATO ever agreed on an armoured offensive as opposed to air interdiction in depth, it would certainly not agree to a _pre-emptive_ land infringement of the IGF. But if the counter-stroke is not pre-emptive, it may not gain enough space and will present the Soviets with precisely the type of encounter battle they seek.

Thus the tide of opinion is beginning to run towards an essentially _defensive_, low profile type of defence, in great depth and making the best use of all types of terrain, including what the U. S. military so aptly calls "urban terrain". "Fortifications", as mentioned earlier, has even become a printable word again. As long as NATO maintains anything like its present posture, the crunch point --

since 1980 a publicly acknowledged one -- is in fact how to hold the main Warsaw Pact thrust on the North German Plain.

There is one other fundamental that, I suggest, few people on either side of the Iron Curtain would dispute. And that is that every defence force needs to build up its special forces as far as it can, without dilution of quality, and to make certain that the skills and hardware to put them in by air or sea exist.

To draw together everything I've been saying, special forces apart, one can foresee a need for five types of brigade: parachute, helicopter, infantry, mechanized and light. The parachute brigade is standard; there will always be a need to reach out farther and faster than helicopters can go, and to secure helicopter airheads. For the helicopter brigade the United States air cavalry regiment and the Soviet airborne assault brigade offer remarkably good models. The standard, or line, infantry brigade is also normal. Just as infantry battalions acquired trucks in the Thirties, they will, I'm sure, be equipped with, or at least be scaled for a light workhorse APC such as the Canadian "Grizzly" or the American M113A1. But they will not be mechanized battalions in the sense of fighting the manoeuvre battle with armour. For their anti-tank capability, these standard battalions would, I suggest, have rocket launchers as section weapons, manportable ATGW as company weapons, and -- like the Soviet battalion -- a support company consisting entirely of mortars.

As for the mechanized brigade, I would suggest that given a tank successor of about 15 tonnes and "double articulation", the entire armoured vehicle requirements for it and its supporting arms could be met from a single set of hull sections and basic major assemblies. Three sub-families -- short, standard and extended -- in tracked and wheeled form would roughly correspond to the reconnaissance, main manoeuvre force, and combat support roles; in round figures they would weigh ten, fifteen and twenty-five tonnes, respectively. The point is, of course, that commoning the tank with the rest of the AFV family would have a radical effect on force structure, allowing the combat arm to be streamlined.

I see the mechanized brigade being composed of composite -- that is combined tank/infantry -- battalions. In fact, composite battalions, with two tank and two mechanized infantry companies, have existed for quite a while in the French and Swedish armies. With the tank and the infantry section vehicle coming from the same family, the arguments against them are mainly tribal or institutional.

A far more interesting notion is the integrated company and platoon (Fig. 6-24). There seems to be a steady and growing tide of opinion in every army, including -- as I mentioned -- the Soviet Army, that the most powerful and flexible mechanized force is a balanced one. The crunch point here is that the roles of "in-house" infantry (to borrow the German expression) are not to hold ground, but to get the tanks forward in close country and to relieve the tanks of some of the fire tasks needed to deal with the direct fire threat spectrum. By the same token, the tanks sustain the infantry element's mobility in open country. In my view, this calls for the most intimate cooperation down to platoon level. But there are two aspects to this -- tactics on the one hand, and manning and training on the other.

Taking tactics first, there will be occasions -- though I find them remarkably hard to name -- when a tank-heavy force is required. Certainly there will be situations -- blocking positions, for instance -- which call for concentrations of up to a company's worth of mechanized infantry; and even times when this infantry must operate for a period divorced from its own vehicles as well as the

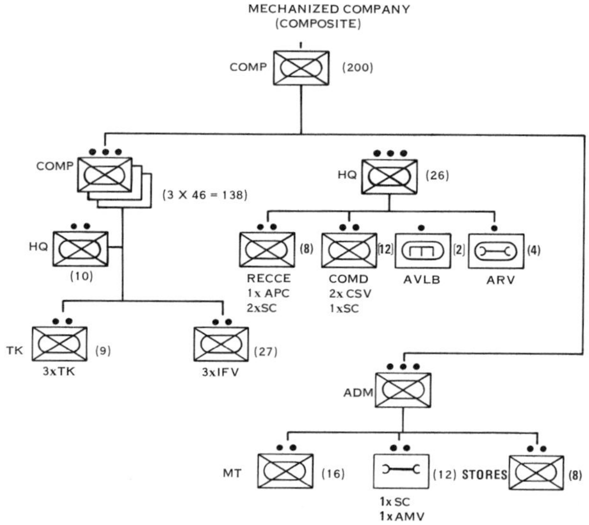

Composite Mechanized Company
Fig. 6-24

tanks. But I or any other tank man who has served much in infantry formations, can tell you that it is just as easy to adjust balance by detaching as by attaching. So surely it is logical to structure the mechanized unit for the most frequent and most important combat situations. That, I suggest, is the balanced unit.[5]

To reinforce this view, a common family of equipment stands the conventional arguments about manning and training on their heads. With the kind of tanks I have depicted, even the most highly educated and technologically aware countries are going to be pushed to find enough suitable men for the tank commander's and deputy commander's seats from within their army's manpower pool. On top of this, reducing the crew to three and requiring a second member of NCO quality is going to aggravate the "all chiefs and no Indians" problem that has bedevilled armoured units in the field and in barracks for as long as I can recall. If, on the other hand, you have a large platoon of forty-five or fifty with a mounted and a dismounted -- or rather dismountable -- element, it should be possible to select and bring on the nine high-grade crewmen needed.

On the training side, one can link this requirement to the need for intimate cooperation. As indicated in Fig. 6-25, everybody starts as a "squaddie". Some will find their level as drivers. The more promising ones, who will probably need to be long-service soldiers, can graduate through deputy section commander to deputy vehicle commander, thence to vehicle commander or section commander and so on.

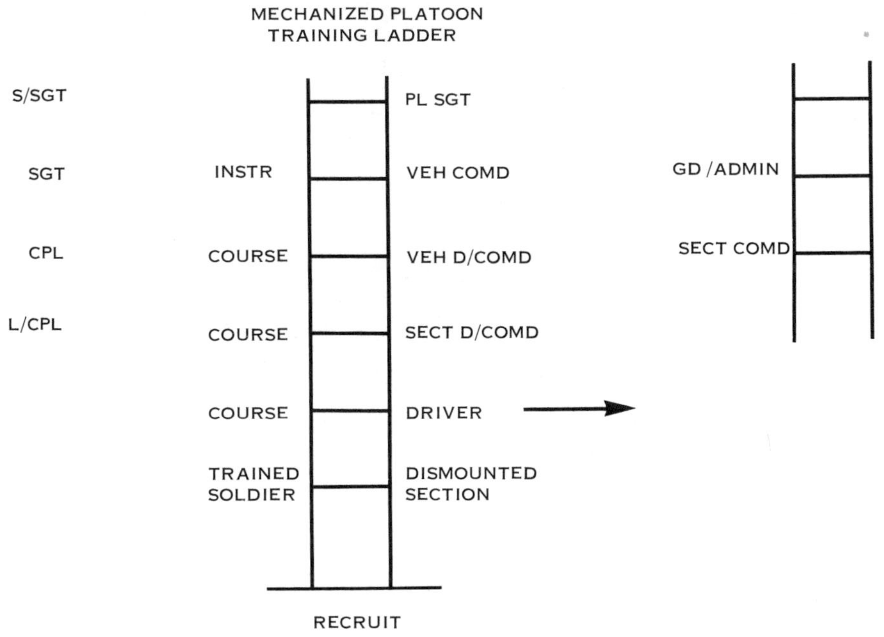

Mechanized Platoon Training Ladder
Fig. 6-25

Before turning to the light brigade, I would like to mention artillery support. And I hope the gunners among you will allow me to refer to the unit of three batteries that we now call a regiment as a battalion. The parachute and helicopter brigades will evidently need tailor-made mixed artillery units -- perhaps a five-battery regiment -- organic to them. I haven't changed my tie, but I'm convinced that, in the long run, each infantry and mechanized brigade will require an appropriate regiment of three battalions, nine batteries that is, to support it. This is why both the combat arm and the artillery must be streamlined.

You will see that two of the types of brigades I have postulated are, if you ignore their particular means of mobility, half-way houses between special forces and the relatively conventional infantry and mechanized brigades. Clearly the character and training of the dismountable elements in the two air-mobile brigades would fit them very well for quasi-guerrilla operations. But I want to put to you the view that the importance of quasi-guerrilla operations is so great, and growing so fast,[6] that this half-way house needs a third storey without any expensive operational mobility organic to it. This I have called the <u>light brigade</u>, not -- I assure you --

after the Battle of Balaclava, but to link it firmly to the light infantry tradition founded by Frederick the Great and developed by General Sir John Moore. The light battalion, as Fig. 6-26 indicates, is about the simplest organization of which you can think. It has no support company. This manpower goes to the rifle platoons, making them up to, say, forty, so that they can field three sections with a battle strength of nine -- three squads of three. Platoon headquarters would have enough fat to provide a squad as a bodyguard for the OP parties and sapper detachments from the supporting units.

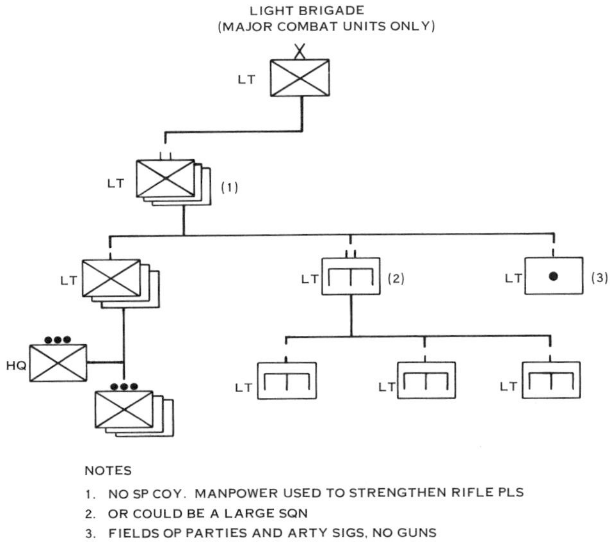

Light Brigade
Fig. 6-26

I have deliberately put the organizational cart before the operational horse, because I want to present a concept that one might call the "deep containment defence" -- or more graphically "hammer, anvil, and net". A fourth element would be associated with it, namely, a permanent and rather sophisticated frontier obstacle belt. Fig. 6-27 shows the concept applied to a piece of ground -- the pass between the Eggegebirge and the Briloner Hoehen with the autobahn running northwest from Kassel. The symbol just under the word "Egge" is the mechanized brigade. To the anvil and the hammer I shall give short shrift. The anvil is a strong positional ("framework" or "lattice") defence, at a density above the limit for acceptable nuclear risk. The hammer's primary role, in this case the mechanized brigade, is to deliver a tactical counter-stroke round or through the anvil with the aim of destroying the enemy in front of it.

The net, shown with two layers in front to exploit the forests and the Weser, is deployed at about one tenth of the density of the anvil

-- a battalion in a rather larger area than a brigade. And this is a maximum if conventional infantry ideas like holding or controlling ground are to be avoided.

The net is there to get the best use out of this terrain by canalizing the attacker, and by administering the kind of repeated pinpricks best calculated to disrupt the Soviet advance. Its ultimate success would be to force the Pact to move up massed infantry, dismount it, and clear through. This would have a dramatic effect on tempo and on the forward movement of mobile troops and ammunition.

Before elaborating on how the net fights, I would like to cover how it extricates itself to fight again. I have discussed this a lot within the British Army, and I reckon the only way to get the net elements out is on their feet. If this is to be an organized withdrawal, as opposed to an escape and evasion exercise, the net must be so disposed that its elements can progressively withdraw onto and through one another.

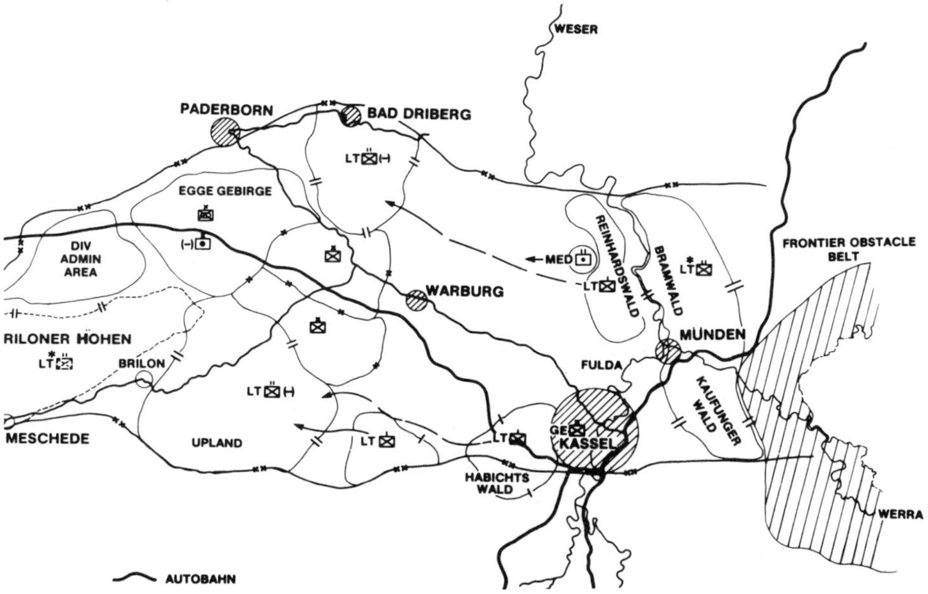

Hammer, Anvil, and Net
Fig. 6-27

This net battle itself has to be a platoon one. Typically, the rather larger platoon I postulated could deploy six tank-hunting squads of three, and one artillery observation party (OP) protection squad, as well as using its headquarters and one section to mount an ambush. The density might average one element per two square kilometres (two map squares, that is), so that a platoon could have sound depth on a frontage of some two kilometres -- making about twenty kilometres for a battalion. The detailed roles are fairly self-evident: tank hunting; calling down fire and air attack; blocking, mining, and booby-trapping; ambushes in conjunction with fire support and mobility denial at key nodes and defiles. These light battalions would evidently need special training, including elementary sapper tasks; above all, they would need a special attitude of mind and the self-confidence to fight independently in very small squads.

The opposition is first checked by the obstacle belt, then slowed down, thinned and canalized by the net. On his side both of the obstacle belt and of the forward net, he concertinas and presents air and rocket artillery targets. He has difficulty in building up to attack the anvil; and is struck either at this stage, or after an unsuccessful attack, by the hammer. With luck it may be possible to restore the position. Otherwise there should have been time to reform the net behind the main defended area, and for the mechanized brigade to cover the anvil force out through it.

I shall now conclude with one final thought. You will probably agree that one must retain some kind of tactical offensive capability in the shape of the mechanized brigade. But given a certain total strength of infantry, might it not be better, as the power of indirect fire systems on the one hand and manportable weapons on the other continues to mount, to use the lot in the quasi-guerrilla role?

Editors' Notes

1 United States Army Tank and Automotive Command.

2 See R.M. Ogorkiewicz, "Brandt Gun-Mortars", International Defense Review, No.2 (1982), 201-204; and "Satory IX: The French Army Equipment Exhibition", National Defense, No 390 (September, 1983), 53.

3 Richard Simpkin, "Hammer, Anvil and Net", British Army Review, No 72 (December, 1982), 11-21.

4 See also General Frederick J. Kroesen, "The Short Cut", NATO's Fifteen Nations, Special Issue 2 (1981), 40-44. On the "standing start" attack see P.H. Vigor, "Soviet Army Wave Attack Philosophy," International Defense Review, No 1 (1979), 43-46 and John Despres, Lilita Dzirkals, and Barton Whaley, The Manchurian Model for Soviet Strategy (Santa Monica: Rand, 1976).

5 During questioning Brigadier Simpkin reiterated that for high-intensity operations on the sophisticated battlefield "a balanced tank infantry force is the most powerful and flexible for most combat situations". He at the same time assessed a tank infantry organization ratio of 2:1, either way, as "very awkward" at any tactical level. For a more detailed examination of the question of tactical balance see Simpkin, Mechanized Infantry, 19-20, 54-55, 58, and 68-69. The defender of Cassino in the Second World War, General Frido von Senger und Etterlin, thought the panzer grenadier rather than panzer division "had the correct balance" and was hence "more suitably organized for all kinds of use in attack and defence than other types of divisions". General Frido von Senger und Etterlin, Neither Fear Nor Hope (London: Macdonalds, 1963), 225.

6 Fuller deduced from his study of history, and the Peninsular and Franco-Prussian wars in particular, "that guerilla wars may run concurrently with great ones." Sceptics need only note the seriousness with which the Allied High Command viewed the German Werwolf threat in the Second World War. See Fuller, Lectures on F.S.R. II (London: Sifton Praed, 1931), 3; and "Is Europe Going Guerilla?", reprinted in Thunderbolts (London: Skeffington, 1946), 111-114. The prescient Miksche also proposed a "guerilla zone" in his "web defence" tactical concept. Blitzkrieg, 204-214.

CHAPTER SEVEN: AT CLOSE QUARTERS

THE LAST 300 YARDS

by
Major D. A. Collett, MC
3rd Battalion, The Parachute Regiment
Aldershot, Hampshire

You have probably all read and heard about the Falklands from professional writers and commentators so it will, therefore, come as no surprise to you if I say that as far as the modern battlefield is concerned, this campaign was not unique or revolutionary in any way, be it in numbers of men fighting, equipment, or tactical lessons learned. What came across to us on the ground, of course, was that the campaign seemed to take on the form of a prolonged exercise, an exercise without "umpires". It was only when men started dropping and not getting up did one suddenly realize that this was indeed the real thing. The aim of my short presentation will be to give you some personal reflections on what it was like, particularly in regard to the last 300 yards of an assault. I shall end with a summary of a few lessons that had to be relearned -- and I would emphasize relearned -- for there were few new lessons from this campaign though many reminders of old ones that had been forgotten. To get to the last 300 yards, however, I need to start at the beginning.

EDITORS' NOTE: The 3rd Battalion, The Parachute Regiment (3 Para) was the Spearhead Battalion and the Leading Parachute Battalion when the Falklands crisis started. The battalion had been part of 8 Field Force, based at Tidworth, since July, 1980. 8 Field Force was designated 5 Infantry Brigade at the beginning of January, 1982. Throughout the Falklands campaign, 3 Para was under command of 3 Commando Brigade.

When the Falklands War broke out, I was commanding "A" Company with 3 Para, having recently returned after an eight year absence. Following the landings at Port San Carlos Settlement by 3 Commando

Brigade (Royal Marines), of which 3 Para was a part, the formation moved out by foot and helicopter to a position west of Stanley. The rocky and boggy terrain, consisting of wide open spaces with hills dispersed across it, was crossed quickly in appalling weather. As Argentinian artillery (had it deployed observers) would have decimated troops moving forward in the open in daylight, most of 3 Para's movement was carried out at night. By evening 31 March, we had moved to within 25 kilometres of Stanley and were in a position in the area of Mount Estancia. Concurrently, 45 Commando was still moving up on foot, 2 Para was resting after taking Goose Green, 40 Commando remained holding the bridgehead at Port San Carlos, and 42 Commando had been helicopter-lifted onto Mount Kent. The brigade had only sporadic contact with the enemy up to this time and, as you can imagine, most of the soldiers we had, marines and paras, were itching to come to grips with the enemy. 5 Infantry Brigade, the other brigade involved in the Falklands, was only just coming ashore at this time and taking no active part in operations.

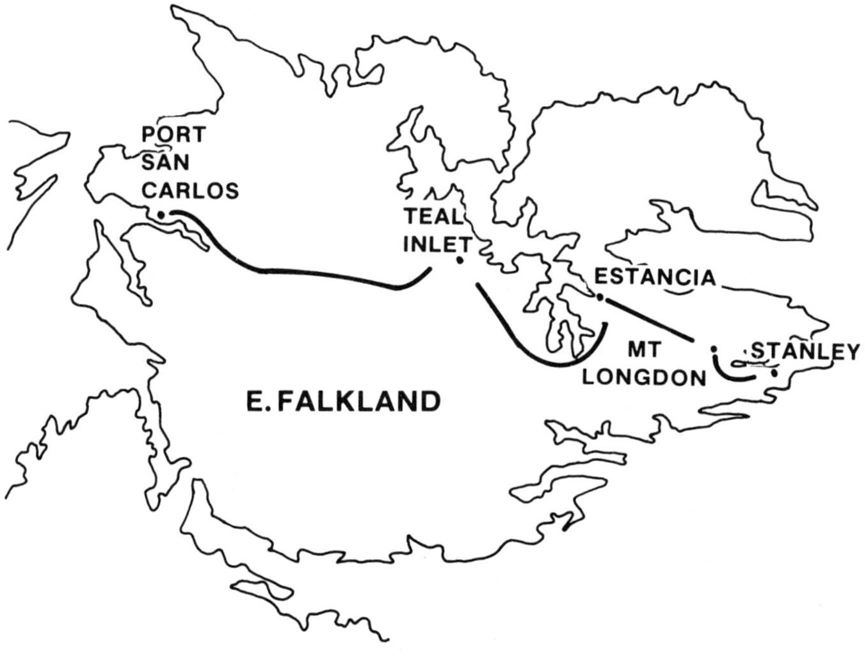

3 Para Advance Port San Carlos to Stanley 21 May - 14 June, 1982
Fig. 7-1

On 3 June, "A" and "B" Companies moved eastward beneath the shoulder of Mount Kent in order to establish secure patrolling bases nearer to Mount Longdon, now known to be our next objective. Mount Kent itself is a high feature, most of the time covered in mist. The terrain around Mount Kent, as most of the Falklands for that matter, contained rock rivers that were quite difficult to traverse on foot. Falling between gaps in the rocks could result in broken legs and wheeled and tracked vehicles could not really get across the rocks. This made life fairly difficult when we were trying to advance, especially at night.

The Argentinians at this stage, so it was determined, were basically positioned on the high ground in a horseshoe shape around Stanley. To some degree they manifested a "wagon train" effect, seeming to gain confidence in numbers. It was estimated they had four or five battalions worth of troops on the high ground, which included Sapper Hill, Mount Tumbledown, Mount Harriet, Two Sisters, Mount Longdon, and Wireless Ridge. The Argentinian troops manning this defensive horseshoe included regular infantry and marine battalions. Supporting them they had 106mm recoiless anti-tank guns (RCLs), 120mm mortars, 105mm pack howitzers and 155mm guns. Around Stanley, they had Panhard armoured cars and LVTP-7s,[1] Argentinian marine armoured vehicles. Besides their mainland air force, they had Puma, Chinook and Huey helicopters, Bell gun-ships, and Pucara anti-insurgency aircraft, slow but effective against unprotected ground troops. In total, the Argentinians had about 10,000 men in and around Stanley. Fortunately, like us, they had poor intelligence and we, the soldiers on the ground, did not know what we were up against.

After moving for about two hours on 3 June, we stopped and dug into some dead ground. My company, having been point company, was at that juncture two kilometres forward of the main battalion. Taking morale into consideration, and the fact we were in dead ground to the enemy who were shelling further back, I decided to stay put. We started digging in at once; however, since we tended to hit water at a depth of two feet, we dug only shell scrapes, next to which we constructed what we termed "peat sangars". In other words, we "built up", and as the days progressed and Argentine shelling became more intense and accurate the peat sangars got bigger and bigger. In this position, which we termed "the bog", we remained for eight days.

In some ways our eight days enforced rest was a god-send, since we had been marching almost non-stop until then. It gave us a chance, first, to wash and shave, for most of the soldiers had not washed or aired their kit since they had started. We also did some training; in dead ground out of sight of the enemy we were able to practice platoon tactics, which gave the troops a chance to actually shake out. We also conducted reconnaissance and active patrolling against Mount Longdon. A further advantage of this situation was that I was able to send platoon commanders and section commanders forward to look at the feature we were going to attack well before we actually attacked. So I was really in a springboard position that was quite good from my point of view.

As the days passed, however, the Argentinians became more active. No fools (and we never underestimated them after initial contact), they had good FOOs and gunners. They realized that a British build-up was taking place as our guns and mortars had begun to pound our future objectives; namely, Two Sisters, Longdon, Mount Harriet and Tumbledown. The Argentinian artillery responded by lobbing shells back and simply conducting harassing shoots by grid squares. They moved a grid square each day to see if they could pick up a target. They also sent a couple of Pucaras against us having first fired white phosphorous (WP). They attacked the area of the WP, just behind my company position, with rockets and machine guns; fortunately, they were well out so we relieved our boredom by firing hundreds of rounds at them. Put off by this weight of fire, they

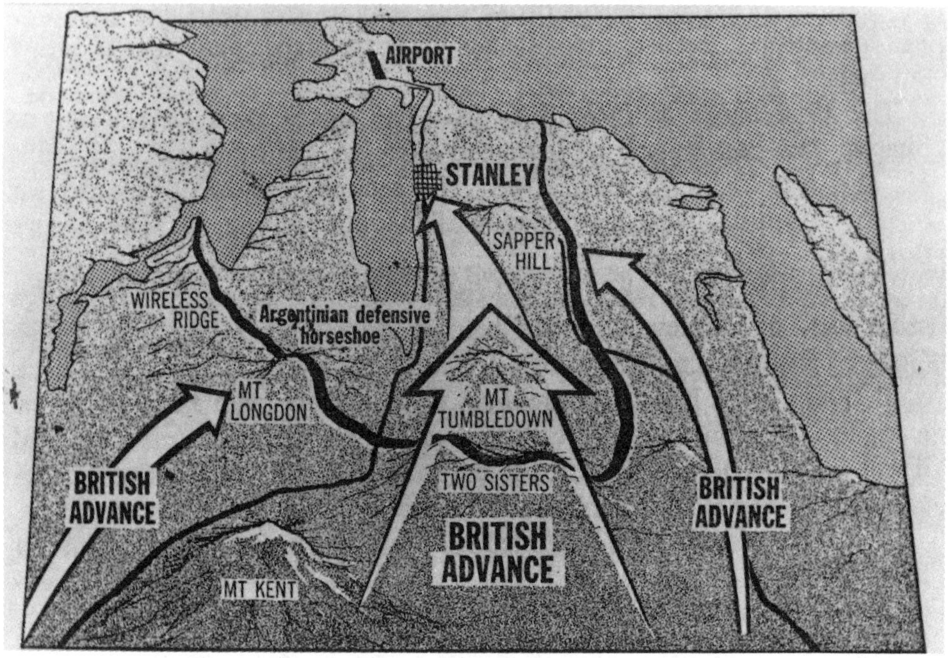

The British Advance on Stanley 12 - 14 June, 1982
Fig. 7-2

left and never came back. The Pucara was basically too slow to risk hostile ground fire.

On 10 June my Commanding Officer gave his orders for the assault onto Mount Longdon. The following day I briefed my platoon commanders in an orders group ("O" Group) plagued by hostile artillery fire. We had enough time, however, to give detailed orders right down to the individual soldier. Basically, my company position was to be the springboard for the attack but because we had taken quite a lot of artillery fire it was decided to build up just back from our position. Civilian landrovers were used as our ammunition and casualty evacuation vehicles. We also employed tractors and landrovers to carry our Milan[2] on an improvised sledge. In fact, we formed a fire base to give covering fire to the advancing troops, using Milan and GPMG(SF).[3] The Milan was used in an anti-bunker or anti-personnel rather than anti-tank role.

EDITORS' NOTE: Maj. Collett graphically described the overall brigade plan which called for a silent night attack with 45 Commando taking Two Sisters and 42 Commando seizing Mount Harriet, south of Two Sisters. Before this, however, 3 Para was to take Mount Longdon. The battalion plan was to secure three summit and ridgeline objectives termed "Fly Half" and "Full Back" (the Longdon summit) and "Wing Forward" (a ridge to the northeast of the summit). "B" Company was to tackle "Fly Half" and "Full Back", while "A" Company seized "Wing Forward". "C" Company was in reserve. Fire Support Teams (Milan and GPMG (SF)) were to move initially to "Free Kick", to the west of Longdon, under Support Company control for subsequent deployment as required. The battalion start line ran along the stream running south from Furze Bush pass. Mount Longdon was held by the Argentinian 7th Infantry Regiment, reinforced by specialist elements and snipers. "Full Back", to the eastern side of the 650 foot height, was a ridge held by Argentinian machine gun and sniper units equipped with extremely effective night observation devices.

After last light on the evening of 11 June, we moved off from my position which was, as I said, the battalion start point. I moved my company in columns of platoons as it gave me greater control. I do not know whether it is normal for all infantry, but in the Para we tend usually to move in long airborne snakes, the leading man in single file walking while the men at the back are doubling; I broke this down into columns with each platoon as part of the column. Again, it gave me better control; it meant we were not strung out.

We moved up to the battalion start line led by a local Falklander; in fact, an ex-chief of police, who went into battle and remained with us the whole time until we reached Stanley. We arrived 20 minutes early, which was fortunate as it gave us time to get our bearings. At midnight we strung out in open formation and moved across the start line. I had two platoons forward (1 Platoon left and 2 Platoon right), my company tactical headquarters group in the middle, 3 Platoon right rear and Company Headquarters Support Group left rear. To facilitate my command of the platoons, I was carrying a Clansmen 349 radio with a throat mike, which I found enabled me to talk directly to platoon commanders without having to go through a radio operator. To control the battle from a reasonable distance is preferable to committing yourself by going too far forward; such action, though very brave, actually risks endangering all your troops, for if a commander gets shot someone else will not only have to assume command, he may well have to start all over again.

At this point let me say a couple of words on our organization and equipment. My two forward platoons were up-to-strength, whereas my third platoon, having suffered eight casualties earlier, was understrength. In addition, since terrain conditions would not permit the use of either our anti-tank Wombats[4] or vehicles, I had been given some members of the anti-tank and motor transport platoons. These I formed into a company support element (which we do not normally have in the British Army), using 84mm Carl Gustav anti-tank weapons and GPMGs, under command of my second-in-command. Each of my riflemen carried about 100 pounds weight, that is, say, all up weight from standing in the nude to having all their kit on with weapon. Radio operators and heavy weapons men carried even more. The individual soldiers' dress and equipment included the following: windproof smock and trousers secondhand, DMS ("directly molded sole") boots (a lot of you may have heard about the fact we suffered from bad feet; I would say it was not due to bad soldiering, but rather to poor boots that were leaky and cold), puttees, helmet, then rifle or sub-machine gun (SMG), five magazines and 50-round bandolier, three grenades (smoke, WP or L2 fragmentation), a 66mm LAW (because we didn't have M79 grenade launchers, we used the light anti-tank weapon (LAW) as an anti-personnel, anti- bunker weapon), webbing, digging tool, water, rations, quilted suit, spare socks and poncho roll. The quilted suit was a substitute for a sleeping bag (we found the sleeping bag too bulky to carry on our webbing, so the quilted suit, wrapped in the poncho roll, did in its stead).

I am sure you will agree that the foregoing items constitute much too heavy a load, especially for close quarter battle. In fact, during the final assault, all my platoons ditched their equipment and fought in smocks with ammunition in their pockets. We collected the equipment up afterwards. Incidentally, every man in my company also carried 100 rounds of link ammunition for the GPMG; this was in addition to that carried by GPMG crews. It is my personal belief that it is the GPMG that wins a battle; the self-loading rifle (SLR) is a superfluous weapon, useful mainly for self-defence or for clearing positions.

Our attack was originally planned to be silent but we had a battery on call in the event that surprise was not achieved. My objective was initially to seize "Wing Forward", while "B" Company moved off further south from me to tackle "Fly Half" and "Full Back". Once they had taken these two positions, I was to move round and take Wireless Ridge. "C" Company were in the reserve throughout.

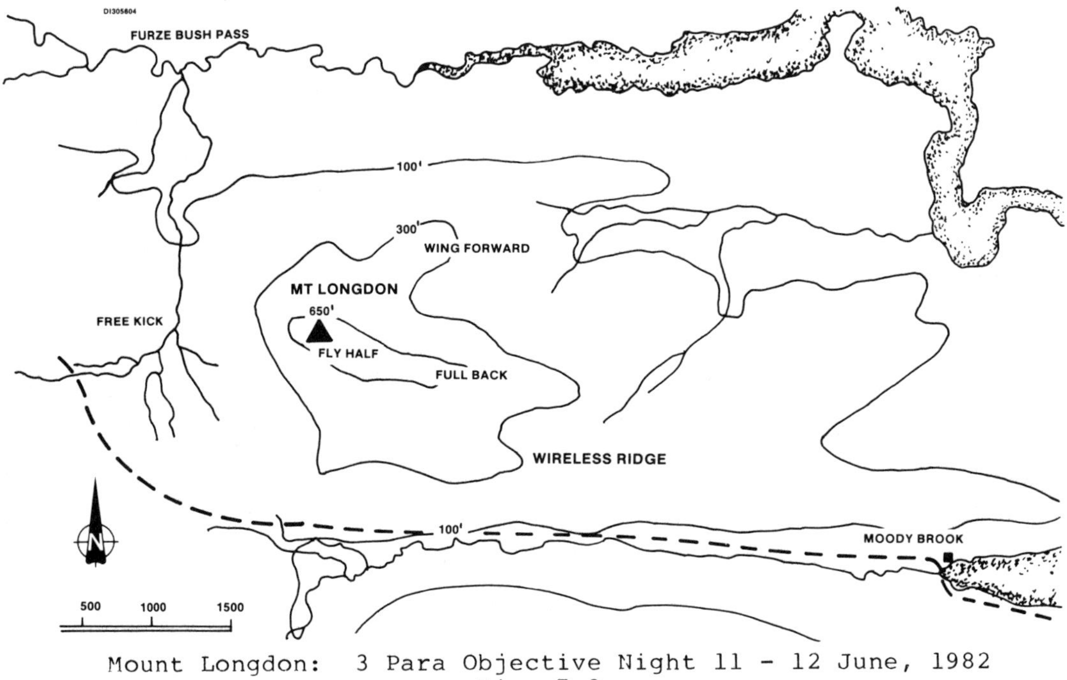

Mount Longdon: 3 Para Objective Night 11 - 12 June, 1982
Fig. 7-3

After sorting ourselves out (those in the infantry will know what I mean), we advanced toward "Wing Forward". We moved up to just below the crest line of "Wing Forward" where we were out of sight of

Longdon summit. Finding no enemy on the position, I went to ground there initially because "B" Company had been slightly delayed in their advance from the start line. The moon, which was going to be full, was just coming up but fortunately it was slow in doing so. We were, in any case, still in the shadow of the mountain which gave us useful cover. Suddenly there was a dull explosion to the south which I knew (having heard them previously in other places) was an anti-personnel mine exploding. Shortly after, our artillery opened fire on Longdon and, soon after that, small arms fire erupted on the western end of Longdon around where "B" Company was moving. My Commanding Officer ordered me to advance parallel with "B" Company which I did.

As "A" Company came over the crest of the hill, we began to receive small arms fire from Longdon, the Argentinians having obviously woken up by this stage. Surprise was now lost and they began to fire their artillery defensive fire (DF) tasks. Fortunately, these had been preregistered to our rear, but soon they started to put up illumination. Ultimately, they fired RCLs, 120mm mortars and pack howitzers, making the night very noisy indeed. We consequently dug ourselves into some peat mounds and waited for "B" Company to clear their positions so we could move round. We at the same time tried to locate the source of Argentinian small arms fire. We discovered later that they were using night scopes and passive night goggles (PNGs), brand new ones I might add, which accounted for their very accurate sniper fire. We were actually pinned down quite heavily and I had two casualties shot straight through the head. Our antiquated IWS, the image-intensifier weapons sight with which we were equipped, was just not up to the ranges at which they were firing. However, not being very familiar with their kit, the Argentinians had the active infra red turned on in some of their equipment, so with our IWS we could actually pinpoint their heavy weapons pits along the side of the hill. Unfortunately, due to the fragmented nature of the battle, the Officer Commanding "B" Company had asked me not to start firing across for fear of hitting his own men. As it was essentially a section battleground, he wasn't exactly certain where his lead elements were.

After some four hours of fighting in which heavy casualties were incurred, "B" Company had almost fought to a standstill. I was consequently ordered to withdraw from my position, move through "B" Company and carry on the advance down the backbone of Longdon. As I went forward on reconnaissance, it became apparent from the debris and number of dead and wounded on both sides that there had been quite a fight. The enemy had dug themselves deep into the rocks with .50 calibre heavy machine gun (HMG) and GPMG interlocking arcs of fire and it was obvious we were going to have to take them out position by position using grenades and bayonets. At the time, artillery was being brought down almost on top of our own men to try to dislodge them. Once I took over from "B" Company, we centralized our equipment and I briefed platoon commanders.

"A" Company had basically to advance over rocky, relatively open, ground that was covered by automatic fire, the crest and vicinity of the ridge to our front was also covered by continuous machine gun fire. I, therefore, sent the FOO and my Headquarters Support Group forward with GPMGs to provide covering fire. I then started feeding men over the ridge one at a time, each in turn working his way forward going from cover to cover, the next man going over crawling to avoid detection.[5] Gradually, as the first few worked forward clearing positions, more men would follow. We put a section over in this manner, then a platoon, and finally two platoons. Dawn was just coming up, in fact, when the enemy seeing that we had enough men over to clear the position, decided to break and pull back. Fortunately, a mist had developed by this time. This was indeed a god-send as it shrouded us while we cleared through the rather extensive position (we actually found Argentinians hiding deep in the rocks even after we had gone firm). The mist enabled us to work through and clear the position without having too much flak thrown at us; it also caused me to sight my company position on a semi-reverse slope! In other words, I sighted it on a side of the hill that, when the mist lifted, was exposed to Wireless Ridge.

EDITORS' NOTE: The men of 3 Para had been in action for six hours by the time "Fly Half" was taken; "Full Back" took a further four hours.

Tumbledown Mountain, which lay to our right, was full of Argentinians. Fortunately, thanks to the mist, we had already reorganized and commenced the difficult task of casualty evacuation which could not be done by air. Looking after the dead and prisoners of war, of course, proved a continuing problem. Not surprisingly, as soon as the mist cleared, Argentinian FOOs (who were, as I said, quite good) started to shell us. This shelling, which continued throughout the day, proved a bit nerve wracking because they were using 155mm guns from Stanley. By this time, however, we had become spectators of a sort. Mount Harriet had been successfully secured by 42 Commando, and Two Sisters by 45 Commando; 2 Para subsequently captured Wireless Ridge, the Scots Guards took Tumbledown, and the Gurkhas claimed Mount William. Shortly afterwards, as we were sitting watching, the Argentinian army suddenly broke.

Within two hours, it was all over.

THE MARINE APPROACH

by

Lieutenant-Colonel Frank M. Izenour and
Major Larry Livingston
United States Marine Corps
Development and Education Command
Quantico, Virginia

According to Ernie Pyle, writing in Okinawa in 1945, a Marine's concept of a perfect battle is to have other Marines on his right and left flanks, Marine aircraft overhead, and Marine artillery backing him up. Current organization for effecting this in war, of course, is the Marine air-ground task force (MAGTF) which is likely to be committed to seize or occupy a piece of key terrain from the sea. The component parts of a MAGTF are: a command element, a ground combat element, an air combat element, and a combat service support element. All MAGTFs are structured in this manner regardless of size or level.

A Marine amphibious unit (MAU) has a battalion landing team as a ground combat element. The battalion landing team is a task organized battalion with a battery in direct support, a truck company, a platoon of tanks, a platoon of engineers, and a platoon of "amtracs" or amphibious tractors. It is supported by the other two-thirds of the air-ground team, the air combat element composite squadron and the MAU service support group. A MAU is commanded by a colonel, the battalion landing team by a lieutenant-colonel, the composite squadron by a lieutenant-colonel and, normally, a major commands the MAU service support group. This is the organization currently deployed in Beruit, a force of about 1800.

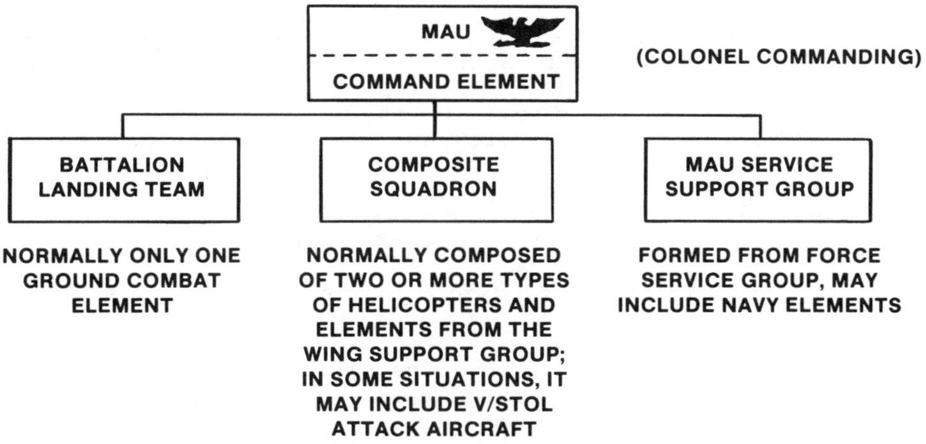

Marine Amphibious Unit Organization
Fig. 7-4

At the next level is a Marine amphibious brigade (MAB). Usually commanded by a brigadier-general, The MAB has a regimental landing team as a ground element, a provisional Marine amphibious group as an air element, and a brigade service support group. This organization can be as large as 16,000, and is probably the first level of a viable fighting force. Above this is the Marine amphibious force (MAF), which could comprise several reinforced divisions, two or more Marine aircraft wings, and an augmented force service support group. It is important to understand that Marine organizations are driven by amphibious lift. In other words, if you wished to have a mechanized amphibious force you would drive your requirement for shipping drastically upwards, after which you would still have to figure out how to fight that organization across a beach. We feel that an infantry organization is the most viable organization to cross a beach and clear it -- the man with a rifle, a hand grenade, and a shovel. You build up from zero combat power as fast as you can and push inland. It is a one way trip -- no place to go, except forward.

EDITORS' NOTE: At this point Major Livingston presented a brief history of the United States Marine Corps from 10 November, 1775, and before. He stressed the essential naval infantry and expeditionary nature of the USMC, making the particular point that rifle marksmanship to 500 metres is still taught. He further intimated that the unique nature of amphibious forces (there is, as mentioned, only one way to get across an enemy-held beach) reinforced the offensive spirit in Marines. At the same time, the historical record demonstrated the four-fold mission of the USMC:

a. to conduct amphibious assaults and subsequent operations ashore;

b. to conduct joint-service operations in conjunction with and in coordination with sister services;

c. to conduct combined arms operations in the prosecution of a land campaign; and

d. to conduct operations as the President may direct.

The fact is, nonetheless, that the US Marines remain the foremost practioners of the art of amphibious operations in the world. No other army, marine force, or naval infantry possesses such an awesome and sustainable air-ground capability for forcible entry over hostile shores from the sea. That the Marines also remain a combined arms model, often but a "raft of sea soldiers in an ocean of Army," is further reason to seriously reflect upon the rationale that has affected their basic force structure.

It is interesting to note that when we built our Marine organizations we started at the bottom level. Lieutenant-Colonels Evans F. Carlson and Samuel B. Griffiths II, two remarkable men, exerted an unprecedented influence on the Marine Corps. Carlson spent half a year fighting with the Chinese Communists prior to American entry the Second World War. Based on his philosophy, we organized our squads in 1944 into a triangular structure of three four-man fire teams, each with a Browning automatic rifle (BAR). All four-man teams were identical, each comprising a fire team leader, a rifleman, an automatic rifleman, and an assistant automatic rifleman to carry ammunition. No man in the Marine Corps was thus ever expected to control more than three men.

What we found out during this period was that it is the number of automatic weapons forward that matters,[6] not the number of men. Manoeuvre was also emphasized at the lowest level, with squad leaders being taught exactly like platoon commanders and almost identical to company commanders. I know from my experience, particularly in the assault where you never regain control until it is over, that the key to success rests in having competent squad and fire team leaders who know what they are doing. The triangular 13-man squad served the Marine Corps well for nearly 40 years.

The infantry battalion was also built on a triangular structure with three rifle companies and a support weapons platoon that included 81mm mortars, water-cooled machine guns, flame throwers, and rocket launchers. The rifle company consisted of three platoons of three squads and a weapons platoon. The latter included lighter crew-served weapons that permitted the massing of firepower at the company level. A Marine attack was thus characterized by a lot of automatic weapons forward and heavy concentrated machine gun fire. Rifle companies assisted each other with enfilade machine gun fire from all positions on the battlefield. Marine tactics traditionally dictated that it was the company commander who kept the machine guns, manoeuvring them as necessary to support forward platoons.

EDITORS' NOTE: In 1961 the M14 replaced the BAR and the M1 Garand rifle. With the adoption of the M79 40mm grenade launcher in 1962, a fourteenth man was added to the Marine squad. Post-Vietnam introduction of the M203 40mm rifle grenade launcher in place of the M79 enabled the squad to again be reduced to 13 Marines. By this time, of course, a fourth rifle company had been added to the Marine battalion and the weapons company deleted.

The latest Marine Corps programmed reorganization, first announced in 1979, calls for restructuring the infantry battalion into a headquarters and service company, weapons company, and three rifle companies (with a fourth rifle company planned in event of mobilization). The weapons company now includes an HMG section equipped with eight Mark 19 40mm automatic grenade launchers (which give good area coverage and can knock out a BMP at under 2000 metres) and eight M2HB .50 calibre HMGs; the section consists of four squads, each containing two two-man teams (a team can only man one of its two weapons at any particular moment, however).[7] It also includes eight 81mm mortars and 32 Dragon 90mm MAW launchers.[8] The emphasis on anti-tank capability is a direct result of the realization that our traditional Marine structure did not sufficiently address the modern armoured threat.

The rifle company still has three platoons and a weapons platoon, the latter comprising a machine gun section, a mortar section, and assault squad. The three sections of the weapons platoon are equipped with six M60 machine guns and three new light-weight M224 60mm mortars (HE range 3,450 metres), the replacement for the aging M72 LAW, which could possibly be the 84mm Swedish AT4 currently being tested (HEAT range 300 metres plus), will be issued as needed (there are no TOWs or Dragons in the rifle company). The rifle squad restructure has, of course, become the focal point of controversy. The 13-man squad, of three four-man fire teams, is being replaced by an 11-man squad of two five-man fire teams, each equipped with an M249 (Minimi) machine gun or squad automatic weapon (SAW) firing 5.56 rounds out to 800 metres from a magazine or belt. With this weapon the squad again acquires a specially designed automatic weapon, something it has not had since the retirement of the BAR. Platoon strength thus goes down to one officer and 35 men.

While you may think that sounds like a lot of men, to us it is not. I personally have reservations about this organization; not about the three rifle squads, but about the two fire teams with five men in each. For the first time in 40 years we have made the fire team leader, the lowest ranking NCO in the squad, command more soldiers directly than anyone in a Marine division. We're getting away from our concept of fire elements forward, and we're now starting to deal in men forward. The platoon commander is reduced by one-third in his automatic rifles; the number of men are not reduced substantially, but the useable light automatic teams are cut down by a third. At small unit level the capability for manoeuvre thus moves one level up. What we once expected from our squad leaders, will now be accomplished at platoon level; two elements at squad are just not enough as only three elements can manoeuvre.[9] This, of course, is something so against our philosophy I don't know whether we will be able to live with it. (1988) They could not, and restored the 3FT, 13 man, squad.

MAGTF Doctrine is Over 35 Years Old
Fig. 7-5

The key point is that we have dropped from nine fire elements to six, which is certain to present problems when it comes to closing with an enemy over the last 300 yards. The mission of the Marine rifle squad in the attack, as in most other infantry, is to locate, close with, and destroy the enemy by fire and manoeuvre. We define "fire and manoeuvre" as a technique of advance in which one element of a unit, the manoeuvre element, moves while being supported by another element, a basic fire element ("fire and movement", on the other hand, is a technique of advance by which the manoeuvre element simultaneously fires and moves by bounds). The problem as I see it, is that whereas before two integral automatic rifles covered squad movement, there is now but one. If an envelopment is planned, and we do try to envelop whenever possible (even when our higher organization is attacking frontally), the squad leader has only one automatic rifle to take round. Should such an envelopment element run into trouble, as was often the experience in Vietnam, the manoeuvre is doomed to founder. Simply put, it lacks the automatic firepower necessary to cover a flank threatened by automatic fire; when you give up machine guns, you give up your ability to suppress enemy machine guns that are dug-in or in bunkers. Having been a squad leader, platoon commander, and company commander, I can assure you that the organization one needs when going into the final assault is an organization that has lots of fire elements immediately available. It is these elements working as a team that fend off disruptions to an assault, clear through objectives, and hold against counter-attack.

As far as I am concerned, a squad leader capable of controlling the fire and manoeuvre of three fire teams can run a rifle platoon at night. In the Second World War, triangular squad organization actually proved a fertile tactical training ground for junior leadership. Everyone was versed in the same fire and manoeuvre tactical doctrine from the most basic level up. I fear now that the squad leader will have no option but to go forward frontally. Moreover, not being schooled in manoeuvre, it is unlikely such a squad leader will be able to handle a platoon.

EDITORS' NOTES

1 U.S. Marine amphibian tractor or "amtrac".

2 Franco-German anti-tank guided missile (ATGM) system; with a range of 2000 metres it falls into the medium range anti-armour weapon (MAW) category.

3 General purpose machine gun (sustained fire). It is a 7.62mm air-cooled, belt-fed weapon.

4 British 120mm anti-tank weapon similar to the 106mm RCL.

5 Major Collett stated during question period that at this juncture he urged his troops to proceed cautiously as "there was no point in losing more men". He further stated that action over these last few hundred yards quickly devolved into a "section commander's and soldier's battle". Only they, the men on the ground, could actually judge what had to be done; higher commanders could do little else than await results. The company support group that he had formed did, of course, prove its worth in this situation. He

stressed again, however, that it is "machine guns that win the battle" and that there should be a minimum of "two guns per section" (he added that the rate of consumption of belt-link ammunition was high, though there was no shortage of SLR ball). A lack of HE ammunition for the 2-inch mortar limited the usefulness of another valuable close support weapon (the firing of smoke and illuminating rounds being considered a "waste of time").

6 This is a similar conclusion to that of Major Collett. It was also the discovery of the German army, which eventually developed the "most advanced and effective machine gun of its time," the MG 34. The 45,000 German sections that invaded the USSR in 1941 were each armed with this weapon (the more battle-experienced soon acquired two), which gave them an edge over their Soviet counterparts. The Germans, accordingly, almost always attacked flanks, trying to drive an enemy out by superior firepower. Lieutenant-Colonel Tony Jeapes of the British SAS fighting against Soviet-equipped irregulars in the more recent Dhofar Campaign (1971-76) drew a similar conclusion that is worth repeating: "The effect of even a dozen men armed with AKs alone opening up at you at close quarters is mind-numbing in its violence. It is not the rapid pop-pop-pop of musketry. It is the overwhelming shattering shock of a dozen machine guns (for that is what a fully automatic weapon is) firing at you from the length of a football field. The only way to make any dent upon the enemy is to get as low as possible and fire back with everything you have. Any idea of movement in an area into which six thousand rounds a minute are being directed (12 x 500) is out of the question, yet this is what our soldiers must expect to face in the final one hundred metres of the assault. Under those conditions you do not assault. You manoeuvre into a position from where you can kill the enemy by fire. Not until all resistance has stopped do you move in and then only to pick up the pieces. Even one determined man left alive will wipe out in seconds a section advancing on foot. I have seen it happen.... The fact is that when the fighting was heavy, the SLR was left behind.... Automatic weapons were at a premium... and above all, that wonderful weapon the GPMG.... No doubt it is a curse to lug about the forests of Sennelager and the plains of Soltau, but the GPMG is a killer, the one weapon that will silence the enemy. The AR15 is accurate, light and automatic and the 5.56mm round is fine as far as it goes, but it cannot chew its way through trees, rocks or sandbags the way the GPMG does to get at the enemy behind." Lieutenant-Colonel A.J. Jeapes, "Letter to the Editor," British Army Review, No 63 (December, 1979), 73.

7 The HMG section rates eight of each weapon but can employ only eight -- in any combination -- at one time. The M2HB has a maximum effective range of 1,830 metres, but against armour this range is substantially less. The saboted light armour penetrator (SLAP) round now being developed is designed to improve the armour-piercing capabilities of the weapon. The MK 19 is the better armour killer. It fires high explosive dual purpose (HEDP) rounds that are effective against hard and soft point targets out to 1,650 metres. The MK 19 round takes 17 seconds to travel 2,000 metres. Lieutenant-Colonel R.A. Beeler "New Infantry Battalion Structure: An Overview" and Second-Lieutenant John F. Schmitt, "Heavy MG Section in Crawl Stage," Marine Corps Gazette, No 6 (June, 1983), 18, 21.

8 The U.S. stopped production of Dragons in 1978. Because of its short (1000 metre) range, limited lethality and associated training difficulties it will eventually be replaced by the Infantry man-portable anti-armour/assault weapon system (IMAAWS) or Tankbreaker Anti-tank Missile. Major John P. Gritz, "Light Armoured Vehicle or Light Armored Victims?", Marine Corps Gazette, No 8 (August, 1982), 41-42.

9 In Commonwealth organization the platoon has historically been the basic manoeuvre element or tactical unit. According to a survey completed in conjunction with the Marine infantry battalion study, 47 per cent of field grade and company grade Marines officers polled would prefer to retain the three four-man fire team squad. Major Terrence P. Murray, "Establishing Force Structure and Doctrine", Marine Corps Gazette, No 9 (September, 1983), 62.

CHAPTER EIGHT: INFANTRY COMBAT VEHICLE INSIGHTS

THE BRADLEY ADVANCE

by
Lieutenant-Colonel Michael H. Hansen and
Captain John Carmichael
United States Army Infantry School
Fort Benning, Georgia

The U.S. Army's new Division 86 structure addresses only the heavy aspects of the infantry; it does not address the airborne, air assault, or light infantry at this time in our army. These organizations remain the same and we do not visualize them changing until possibly the late 1980s. The 9th Division is our high technology experimental division. Once our army decides with what this division is going to be equipped, and how it is going to fight, the 25th Division in Hawaii, and the 7th Division in Ford Ord, California, are likely to be similarly organized. These will become our three light divisions.

Change in U.S. Army doctrine has been an evolutionary process and we now adhere to our air-land battle doctrine. I do not want to discuss this, however, because it does not affect essentially how an infantry battalion will fight on the modern battlefield. We face very few changes at battalion level as a result of air-land battle doctrine. The equipment and weapons we are bringing into our tactics are, on the other hand, forcing us to adjust our tactics. Examples of these weapons are the Ml Abrams tank, the M2 Bradley infantry fighting vehicle (BIFV) and a host of other weapons and equipment.

The first BIFV battalion will be fielded at Fort Hood, Texas, and the second battalion will form in Germany in the 3rd Infantry Division. Basically, the Division 86 mechanized (BIFV) battalion will comprize a headquarters company, four line companies of 13 BIFV each, and one anti-tank company consisting of 12 improved TOW vehicles (ITVs). Six 4.2 inch mortars will be grouped at battalion level. All told, there are 108 anti-armour systems within the Division 86 mechanized infantry battalion.

The BIFV itself is a formidable fighting machine. Weighing 22.7 tons, it carries a total of seven anti-tank missiles: two TOW with the launcher; and an additional five in the vehicle, the mix of which, either TOW or Dragon, is based upon the commander's tactical assessment. The BIFV also mounts an M242 25mm Bushmaster chain gun that fires high explosive incendiary tracer (HEIT) and armour piercing discarding sabot (APDS) ammunition. Both the BIFV and its

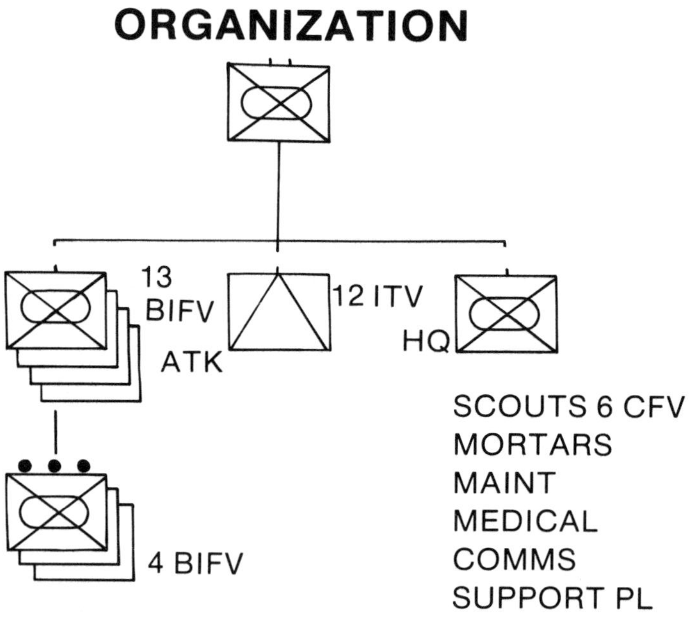

U.S. Army Mechanized Battalion Organization
Fig. 8-1

cavalry fighting vehicle (BCFV) version have a stabilized turret. This permits accurate fire to be delivered on the move. The additional M240C (MAG 58) coaxially mounted 7.62mm machine gun has a cyclic rate of fire from 650 to 950 rounds per minute; it is used for destruction of personnel and suppression of enemy positions. Each BIFV also sports six firing ports through which 5.56mm rifles can be fired from an open bolt position using a 30-round magazine.[1] The eight smoke grenade launchers mounted on the front of the BIFV use red phospherous and in a period of two seconds can create a smoke screen 10 metres high and 70 metres wide that lasts from one to three minutes. The vehicle can also use its exhaust system to generate a smoke screen. A gas evacuation system is additionally included to provide for crew comfort.

The fuel capacity of the BIFV is 175 gallons and it meets the 300 mile cruising range requirement. Its integrated sight unit is a single sight unit used for all the weapons on the turret. With a day and night sight mode, its optics are 4x power and 12x power magnification. The night mode uses a thermal imager. Though the thermal imager is degraded by dense fog, heavy smoke, rain and snow, minor amounts of smoke have no effect at all on the sight. The integrated sight unit can detect vehicle targets up to 3000 metres and personnel targets at somewhere between 1500 and 2000 metres. Incidentally, the new battle dress uniform with which our army is being equipped has infra-red scattering dye which breaks up its thermal signature. When you look at a soldier who is wearing such a uniform through a thermal imager, all you see are his hands, his face and his boots.

I might add at this point, that with the fielding of the BMP during 1967-68 and the German Marder in 1971, it was determined that a requirement existed for a fighting vehicle that would provide infantrymen with the protection and firepower they needed to cope with the envisioned threat. In this regard, we might now take a closer look at the BMP. A most impressive weapons system, the BMP 1 mounts a 73mm smoothbore gun, a 7.62 coaxial machine gun, an AT-3C (Sagger) or AT-4 (Spigot) ATGM. It also has firing ports for assault rifles and light machine guns; an additional firing port also now appears near the driver's position where there was no firing port before.[2] Of even greater concern, the new BMP 81 now mounts a 30mm cannon as opposed to 73mm low-velocity smoothbore. It would appear that the Soviets are trying to pick up our version of an armour-defeating cannon mounted on an armoured fighting vehicle that has a high rate of fire and long engagement range.[3]

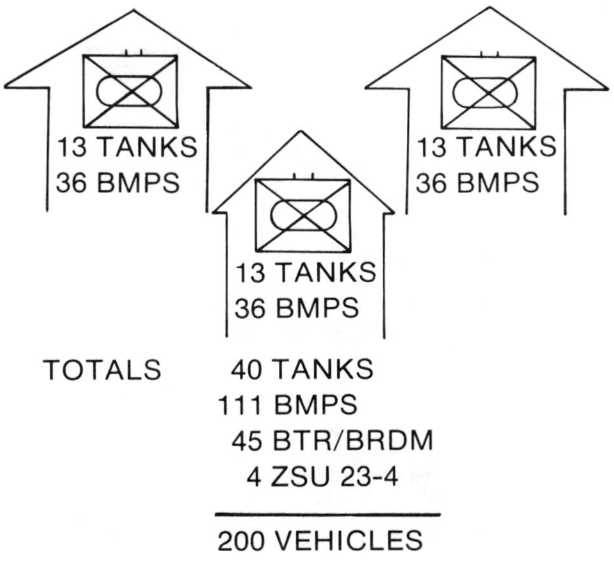

Motor Rifle Regiment
Fig. 8-2

Turning now to a motor rifle regiment, one can see from Fig. 8-2 that we have a total of approximately 200 vehicles. Currently, the Soviets are undergoing the same type of structural reorganization as we are; for example, the incorporation into BMP equipped units of automatic grenade launchers. This rather significant improvement provides a battalion commander with two advantages: first, the three BMPs that carry the automatic grenade launcher platoon are real BMPs (with no turret taken off or anything like that, so the battalion commander automatically has three more BMPs at his disposal for additional firepower and manoeuvreability); and, second, the AGS-17 automatic grenade launcher can provide a very high density of firepower with its 30mm grenades.[4] There are thus approximately 167 armour killing systems within a Soviet motor rifle regiment. While I would not want to create the impression that we are mainly concerned with a mathematical comparison of numbers, I would argue that it is useful to look at what an enemy is likely to field. We should accordingly keep such facts in mind when we briefly examine some of the differences between M113 and M2 equipped units.

Under Division 86 organization the platoon is very similar to Canadian army platoon structure. There are three rifle squads of 10 men each, the platoon leader, his driver, his gunner, his radio operator, and the platoon sergeant. There are thus 35 soldiers in each platoon. We have retained the M-60 machine gun and we have further incorporated the squad automatic weapon (two SAW per squad) that can be fired either with belt ammunition or with a 30-round magazine. So we actually have in each squad three very good automatic weapons. The squad automatic weapon will fire the NATO SS109 round, an improved version of our current 5.56mm ammunition.

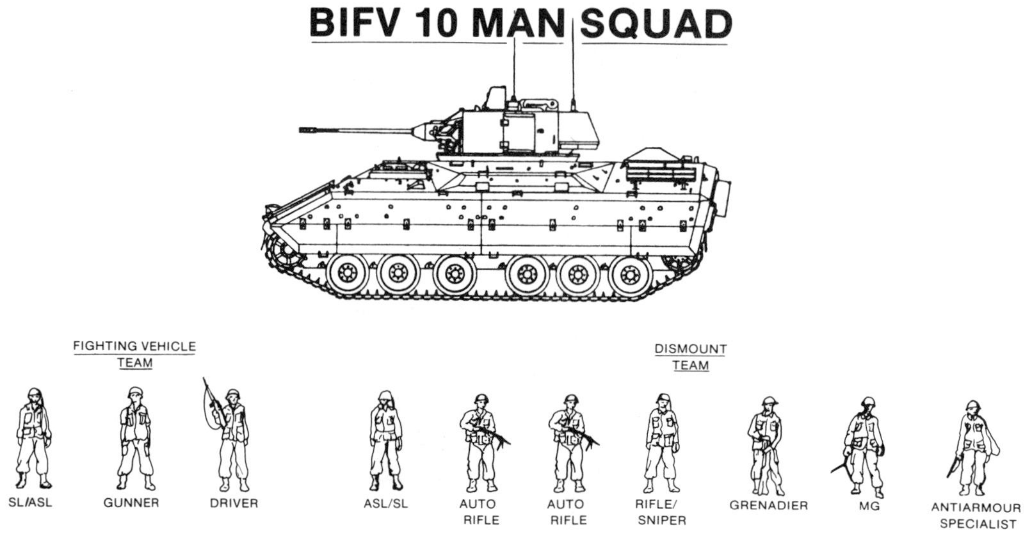

U.S. Army BIFV Squad
Fig. 8-3

The squad itself is broken down into a fighting vehicle element and a dismount element. This configuration can, of course, pose problems of command and control; first, in determining where the squad leader goes, and second, whether the squad is going to fight with the vehicle under the squad leader's control. Alternatively, each squad dismount element could be consolidated at platoon level with fighting vehicle elements remaining separate. A situation may even develop metres. Lieutenant-Colonel English in his book, A Perspective on Infantry,[9] talked about the increasing amount of dispersion that has been required on the battlefield; in tracing the historical evolution of this process, he attributed it mainly to the impact of firepower. I think now, with the inclusion of the IFV in most force structures, that we will see a further extension of this process.

Terrain factors and the enemy are obviously going to determine whether we will deploy into small strong points of perhaps dismount team (squad) size or whether we will occupy more strongly fortified defensive positions possibly up to company size. Smaller elements, even as small as dismount team strong points, are in any case likely to be more dispersed in order to protect themselves from artillery fire and at the same time provide a lattice work within which the mobile fight will take place. The lattice work I allude to is that described by Mr. Messenger or its predecessor, the "archipelago defence" adumbrated in an article called "Armour and Counter-Armour" that appeared in the March-April-May 1944 issues of the Infantry Journal. The latter was written by Major-General J.F.C. Fuller and described a technique used by the Germans on various fronts in the Second World War to deal with armoured mobile attack. It essentially illustrated how the elements of the combined arms team should be used in conjunction with each other to fight the defensive battle in both its static and mobile aspects.

In order to ensure the survivability of dismount elements they should be so placed that the enemy is forced into engaging them with his dismounted infantry. We must never allow our infantry to be caught in a position where they can be engaged outside the maximum effective range of their organic weapons. In fact, it would be suicide to put dismounted infantry in a position where they can be engaged at a range of greater than 1000 metres. This would allow the enemy to pound our infantry from a position of relative impunity that would still be within the effective range of his main weapons.

The BIFV of course, allows us to increase the density and effective range of our anti-armour fire. This will serve to force the enemy to dismount and attack our infantry in an effort to disrupt our defensive framework. This will, in turn, slow his momentum. With the BIFV, the enemy will additionally be forced to employ his direct and indirect fire not just against our tanks and TOW vehicles, but also against our infantry vehicles in order to suppress anti-armour fire to facilitate closure or bypass. With the BIFV we almost double the number of weapons systems that the enemy must overcome to effectively attack our defense. The BIFV, in short, is a vastly improved weapons system that enhances combined arms manoeuvre warfare and gives a mechanized infantry battalion task force the capability to defeat a Soviet motor rifle regiment.

A CANADIAN VIEW

by
Lieutenant-Colonel I. D. Barnes, CD and
Major N. B. Jeffries, CD, Directorate of Land
Combat Development (DLCD), National Defence
Headquarters, Ottawa, Ontario

What follows is a second selected excerpt from the DLCD presentation. Readers are asked to bear in mind that in the course of the Canadian combat development process to this point a number of major branch studies (i.e. an infantry study, armour study, artillery study, engineer study, etc.) were completed and subsequently consolidated into one all-encompassing Systems Study that analyzed tactical concepts from the perspective of seven combat functions (firepower, mobility, protection, intelligence, command and control, communications, and combat service support). The Corps 86 concept sprung from these studies, all of which were directed toward operations within NATO or American-British-Canadian-Australian (ABCA) concepts for the defence in accordance with NATO strategic policy. Organizational structures developed through the studies thus related to the high-intensity battlefield in Europe, focussing particularly on defence against a Warsaw Pact style enemy. This next section briefly discusses the role and organization of Canadian infantry in this operational environment.

Within the overall operational concept for Corps 86, the Canadian combat development conceptual umbrella covering the conduct of land warfare in the 1986 - 1995 time frame, the aim of the defence is to stop and destroy the enemy. It also incorporates both positional and mobile aspects and includes the NATO and ABCA stages of covering force action, main defence battle and counter-moves. The defence is based upon an obstacle, or series of obstacles, covered by infantry in well dug-in ground-holding positions with their anti-armour weapons forming the basic framework. Formation level anti-armour weapons will augment the infantry to cover likely armour approaches. Tanks will additionally be superimposed in critical areas, though the majority will be husbanded in depth as a reserve to counter the unexpected or to exploit success. The emphasis is on stopping and destroying enemy armour using a broad range of anti-armour weapons deployed in depth. Attrition of enemy armour should commence as far out and as early as possible so that the enemy's numerical superiority is reduced before he can engage in the direct fire battle. Obstacles, both natural and artificial, will be employed to stop the enemy or canalize him into killing areas to maximize the effectiveness of fire that can be brought to bear.

Counter-moves should aim to maintain the initiative and to restore the original FEBA when ground is lost. Accordingly, an effective counter-bombardment capability will be required to destroy enemy indirect fire systems. A dedicated reserve will also be necessary if a commander is to block enemy breakthroughs and counter-attacks or capitalize on any situation that can give him the offensive posture needed to re-establish a strong defence or commence offensive operations. Thus, though the anti-armour battle is essentially defensive in nature, it is but part of an overall effort in which the ability to conduct efficient mobile offensive operations remains critical. The balance of positional and mobile elements will, of course, vary with circumstances.

While the successful anti-armour battle will be an all arms land-air team effort, the role of the infantry in it is worth special consideration. Regardless of level of conflict, type of operation, or sophistication of weapons systems in service, opposing forces will eventually close with each other to determine the ultimate outcome of the battle. Though all troops -- whether combat arms, combat support or combat service support -- must be capable of engaging in such close combat, the need to destroy opposing infantry, consolidate an area, or hold ground presents a requirement for troops who are organized, equipped and trained to perform the close combat function as a primary responsibility. This function is fulfilled by infantry, of which various types -- light, airborne, mountain, armoured, and mechanized -- have evolved over the years to meet the variety of requirements dictated by environment, climate, or task. Today, NATO classifies infantry in two categories, distinguished as follows:

> Light (or non-armoured) infantry that is suited for fighting in terrain where tanks cannot deploy (it will fight dismounted, be ground mobile, and capable of helicopter deployment); and,
>
> Heavy (or armoured) infantry that has the qualities of good armoured protection, high mobility, and the ability to fight within a system of armoured combat troops (it will fight either mounted or dismounted).

In combat development studies it was concluded that the most critical capability requirement for the high-intensity battlefield will be for a mechanized infantry organization to form the basis of the all arms battle group. Once such an organization is structured, suitable modifications can be made to adjust for other requirements. Therefore, although Canadian infantry operating as part of Corps 86 will perform tasks associated with both light and heavy infantry, they will be organized, equipped, and designated as mechanized infantry. Unfortunately, we do not have the time to cover all of the study material relevant to this decision. We will, therefore, concentrate on selected points.

The role of mechanized infantry in the Corps 86 context remains "to close with and destroy the enemy". However, the infantry study team in its deliberations reached consensus on an alternative definition that, though not official, perhaps better illustrates the infantry function; namely, "to create the battlefield framework through its physical control of ground and its destruction of the enemy in close combat". The latter definition suggests the need for infantry to combine with other arms to perform its role. Thus, in the majority of operations, mechanized infantry will operate with tanks in formed battle groups or combat teams. These basic groupings (to which can be added other supporting arms and services as required) will continue to be defined as we know them in present convention. The ratio of tanks to infantry in each grouping in any phase of operation will depend primarily on the task allocated, the resources available, the nature of the ground, and the conditions of visibility and weather.

The Canadian mechanized infantry battalion was developed with the bottom-up approach of determining the smallest tactical grouping first. Thus, organization development began with the infantry section. Study determined that to be capable of advancing by fire and movement and destroying an enemy tank killing team, the section should consist of eight dismounted infantrymen divided into two assault fire teams of four. Equipped with an infantry combat vehicle mounting an anti-IFV cannon and manned by a driver and gunner, this ten-man section provides the basic building block for the infantry battalion. It can be controlled by the section commander using voice only and is capable of surviving on the modern high-intensity battlefield. The section can accomplish all anticipated tasks, from controlling vital ground to protecting tanks to destroying the enemy at close quarters in all manner of terrain.

Canadian Infantryman
Fig. 8-7

The infantry platoon was structured on the basis of three sections plus a platoon headquarters. The headquarters, also mounted in an IFV, will include the platoon commander, the second-in-command, a radio operator, vehicle driver, and gunner. A portable medium anti-tank weapon (MAW) and its three-man crew is additionally included with the headquarters. The MAW provides anti-armour protection in dismounted operations away from the combat team and mutual support with other platoons. The platoon, all told, comprizes one officer and 37 other ranks.

At company level, organization development determined that three platoons and a company headquarters were required. This organization has the flexibility of being able to detail a platoon to an armoured squadron in return for a tank troop, while still remaining a viable infantry force within a combat team. The company headquarters contains a fourth MAW, plus the necessary personnel for command and control, ammunition resupply, communications, transport, and internal administration.

Canadian mechanized battalion organization for the high intensity battlefield includes four mechanized rifle companies, a combat support company, and an administrative company. The combat support company contains reconnaissance, mortar, anti-armour, assault pioneer, and headquarters and signals platoons. The mortar platoon has eight extended range 81mm mortars, and the anti-armour platoon 16 dismountable MAWs carried on IFVs. The assault pioneer platoon comprizes five sections (one for each combat team and one responsible for battle group headquarters), and the headquarters and signals platoon provides double-banked communications for the eleven essential communications nets within the battalion. The foregoing combat support elements are considered the minimum required for battalion survival. The administrative company similarly contains the minimum essential administrative and medical support.

In summary, the mechanized infantry battalion is an organization structured for the high intensity battlefield of the 1986 - 1995 timeframe. Of the 995 all ranks, 818 are fighters and only 177 are what some call the administrative tail. This battalion thus has the strength to fight and survive on a battlefield that will make extraordinary demands in casualties and sustained operations. In addition, the battalion is so structured that modification can be made to sub-units (and lower components) to allow for complete, efficient operation in other environments.

THE LAV-25 OPTION

by

Lieutenant-Colonel Barry Montgomery
United States Marine Corps Development and Education Command,
Quantico, Virginia

The introduction of the LAV-25 into the order of battle is an additional subject of contention within the Marine Corps.

The Light Armoured Vehicle Program was initiated in 1980 as an accelerated acquisition for commercially available "off-the-shelf" light armoured vehicles. In mid-1980 the Marine Corps acquired three Cougar (mounting a 76mm gun) and three Grizzly (mounting twin machine guns) armoured combat vehicles from the Canadian Forces for evaluation. As the program progressed, it became a joint USMC-US Army effort with a Marine colonel assigned as the program manager at the U.S. Army Tank-Automotive Command. Four candidate systems were eventually procured and over a seven month period underwent testing at the Marine Corps Air Ground Combat Centre, Twenty-Nine Palms, California. Test results analysis of contract capabilities, and cost projections were used as the basis for selecting the best system.

A five year multi-service production contract has been awarded to General Motors of Canada for the eight-wheeled (8 x 8) LAV-25. During the five year production, the USMC will receive 289 LAV-25s and the U.S. Army 680. A companion research and development contract was awarded for two prototypes each of mortar,[10] anti-tank, command and control, logistics, and recovery common-chassis variants. Prototypes of air defence and assault gun variants were deferred pending further study. All such variants are slated to undergo additional development and testing leading to a production option decision for each version. USMC programmed fiscal year procurement for all LAV models from 1982 through 1989 totals 744 vehicles.

GMC Canada Assault Gun Vehicle for LAV programme fitted with Arrowpointe turret armed with 90 mm Cockerill Mark III gun, 7.62 mm M240 MG and smoke dischargers

GMC Canada Light Assault Vehicle for LAV programme fitted with Arrowpointe turret armed with 25 mm M242 Hughes Chain Gun, 7.62 mm M240 coaxial MG and smoke dischargers

LAV-25
Fig. 8-8

The basic LAV has a gross vehicle weight of 28,400 pounds, eight tons less than the BIFV, which enables it to be carried sling-loaded under a CH-53E Super Stallion helicopter (all LAV variants will also be compatible with the C-130, C-141, C-5A, and landing craft). With a maximum cruising speed of around 62 miles per hour and a cruising range of roughly 415 miles (at a best fuel economy speed of 43.7 mph), it is ideally suited for rapid deployment to Third World areas like the Middle East. Carrying an electrically-powered M242 25mm chain gun, and a coaxially mounted M240C (MAG 58) 7.62mm machine gun, the LAV-25 will carry a squad of nine Marines that includes a driver, gunner, and squad leader. The two-man turret allows the commander to be next to the gunner and the options of staying with the vehicle or going with the dismounted infantry (in which case he must be replaced). The LAV, in short, is intended to be more than just a battle taxi.

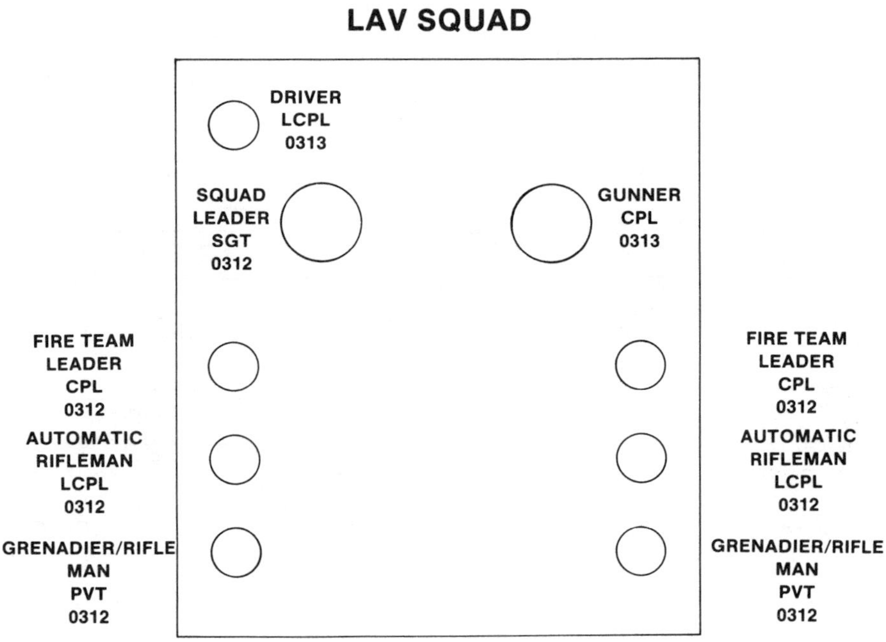

LAV Squad Structure
Fig. 8-9

Concepts and doctrine for the LAV battalion will be refined in an evolutionary manner starting with the first company formed at Twenty-Nine Palms in 1983. A formal Marine Corps LAV School is also being formed at Camp Pendleton, California and the first Marine LAV battalion is slated for activation there in 1984. Another battalion will be formed at Camp Lejeune, North Carolina, in 1985. A year later, a third battalion will be organized at Twenty-Nine Palms, and a reinforced company permanently established on Okinawa.

The Light Armoured Vehicle Battalion (LAVB) will have an establishment of around 1,000 men and will be equipped with 145 LAV vehicles in eight variants. It will include a headquarters and service company, a weapons company, and three LAV companies. The weapons company will have eight 81mm mortars, 15 anti-tank variant TOW vehicles, and 10 air defence variants (as previously stated, the last two versions will not be available before 1986). Each LAV company will comprize an assault gun platoon of nine LAV assault gun[11] vehicles and three infantry platoons, each of four LAV-25s. The three squad vehicles will each be capable of dismounting six men split into two three-man fire teams. The combination of 25mm chain and assault guns (whatever the selection) provide these teams with a substantial amount of direct fire support.

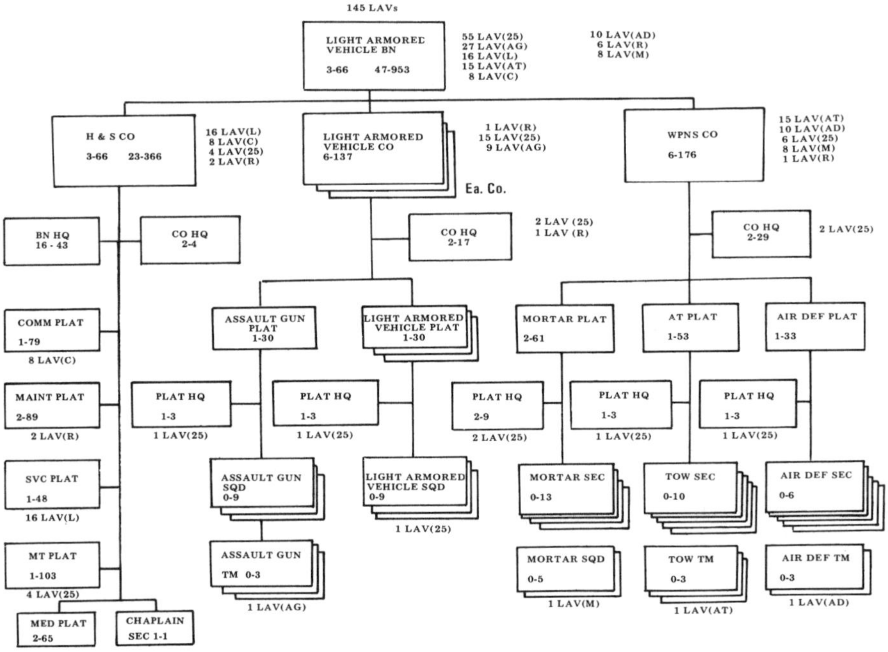

LAV Battalion Organization
Fig. 8-10

The stated role of the LAV battalion is, as part of the Marine division, to conduct recoinnaissance, security, and economy of force missions exploiting the LAVs mobility, agility, and firepower. Destruction of enemy forces will be achieved through its fire and manoeuvre capability. It is not intended, however, that the LAV battalion will form part of the assault echelon in an amphibious landing. It is more likely to be deployed after amphibious operations have taken place and subsequent operations ashore begun. At this point the LAV battalion will probably be given covering force, guard, or screening tasks and possibly even rear area security responsibilities. Helicopter support would of course, greatly increase its survivability, which remains, along with LAV tactical employment, a subject of some controversy within the Marine Corps.[12] It is believed, nonetheless, that the LAV effectively meets the operational requirements of the modern battlefield for greater operational range and tactical mobility. By exploiting the capabilities of the LAV, a MAGTF commander can shape and define the non-linear battlefield of the 1980s. It remains only to work out what specific tactics and techniques must be developed.

EDITORS' NOTES

1 The six firing-port weapons (FPWs) are in addition to the soldier's personal issue M16 rifle and in normal circumstances are not removed from the vehicle. There are no sighting devices in the firing ports and the crew must, therefore, rely on observation of the fall of shot as seen through their vision ports. The FPW can thus only be regarded as an area weapon, since aimed fire is impossible. D.H.C. Jenkins, "Operating the Bradley Infantry Fighting Vehicle," International Defense Review, No 9 (1982), 1204.

2 The BMP 1 carries a crew of three and an eight-man section. The vehicle weighs about 13 tons, is amphibious, and carries four ATGM, forty HEAT rounds of 73mm (effective against armour out to 800 metres), and 2000 rounds for the 7.62 PKT machine gun. It is virtually immune frontally to .50 calibre AP fire. The German Marder, in contrast, carries a crew of three and a seven-man section. It weighs more than 25 tons and is not amphibious. It is armed with a RH 202 20mm cannon, a coaxial 7.62mm machine gun, a remote-controlled rear 7.62mm machine gun, and a Milan ATGM system. Infantryman can fire sub-machine guns through two ball-mount ports on each side.

3 The new BMP 2 (M1981) also mounts an AT-5 Spandrel ATGM (range 100 to 4000 metres, as opposed to Sagger's 500 to 3000). BMP 2 presumably has improved protection and automotive performance.

4 The 40-pound AGS-17 is recoil-operated and can fire grenade-like rounds in either the automatic or semi-automatic mode out to a reported 1500 metres at the rate of 40 to 60 per minute. An anti-armour capable round is also reportedly being readied. Captain L. Rossetto, "Modern Soviet Infantry Weapons," Canadian Defence Quarterly, No 1 (Summer, 1982), 19-20.

5 See Richard Simpkin, "Closing the Survivability Gap"' Armor Magazine, No 6 (November-December, 1981), 19-24.

6 The BDM Corporation, Generals Balck and von Mellenthin on Tactics: Implications for NATO Military Doctrine (19 December, 1980) Director of Net Assessment, Office of the Secretary of Defense and the Defence Nuclear Agency, Contract No. DNA001-78-C-0114.

7 Colonel I. Vorobyov, "Weapons and Tactics", Soviet Military Review, No 4 (April, 1977), 14.

8 Colonel John Boyd, USAF (Retd), who serves as a U.S. Department of Defense consultant and lectures at professional military schools, is the foremost proponent of the idea that an enemy can be defeated most readily by cutting inside his "observation - orientation - decision - action (OODA) cycle." This is accomplished by operating at a faster tempo than one's adversaries. The enemy is unable to cope with the fluid, rapidly changing situation in which he finds himself. Psychologically, the enemy perceives that he has lost control of the situation and is being overwhelmed by events and his own inability to influence the action. The destruction is more psychological in nature than physical. Although the enemy is physically disposed of by selective and local concentrations of forces and firepower at decisive points, he is defeated fundamentally

by destroying his cohesion and his ability to control his forces. Major J.D. Burke, "Manoeuvre Warfare and the MAGTF," Marine Corps Gazette, No 9 (September, 1982), 67-68. During question period Lieutenant-Colonel Hansen pointed out that current U.S. Army doctrine places renewed emphasis on a commander clearly stating his intent. Understanding of the next higher commander's intent, or what he wishes to accomplish, is perceived as being key to quick and effective tactical decision-making at lower levels in fast-moving operations.

9 John A. English, A Perspective on Infantry (New York:Praeger, 1981).

10 Lieutenant-Colonel Montgomery emphasized that the mortar had to be mounted and "be able to shoot 360 degrees from the vehicle". He also expressed serious reservations about the limited range of the 81mm mortar (but 5600 metres, even for the new improved model).

11 The Marines have traditionally employed assault guns for direct fire support of infantry at low-levels, especially against fortified targets.

12 See Major John P. Gritz, "Light Armored Vehicles or Light Armored Victims," Marine Corps Gazette, No 8, (August, 1982), 34-42; Captain Edwin W. Besch, "Mechanizing the Marines," Marine Corps Gazette, No 2 (February, 1983), 17-19; and Major C.F. Hamilton, "LAV and the Amphibious Assault," Marine Corps Gazette, No 10 (October, 1983), 66-73. Major Hamilton argues that in the future Marines are not likely to land into the teeth of enemy defences and sees LAV forces landing simultaneously on the flanks of an amphibious assault.

CHAPTER NINE: MEN AGAINST TANKS

by

Brigadier (Retd) R. E. Simpkin, OBE, MC
ELS Consultant Linguists Limited
Over Deanshaugh, Scotland

This presentation by Brigadier Simpkin was delivered to officers in both Gagetown and Ottawa.

General von Senger und Etterlin, a tank man now Commander-in-Chief of Allied Forces Central Europe, is a very old friend of mine. So I asked him to write an introduction to my first military book. In it, he wrote this: "Even standard national usages are differently interpreted by the various peoples within a nation. A Bavarian does not fight like a man from Hamburg -- nor, I suppose, a Scot like an Englishman." And he finished up: "Finally a word of warning. The tank's main enemy is the tank. Enemy infantry is the opponent of our own infantry ..." and so on.[1] On this subject of like fights like, I remember Joe Backofen, who leads the direct fire weapon systems analysis team at Battelle Columbus Laboratories, writing to me a couple of years back: "It is useless and wrong to pit men against machines, particularly armoured machines." And this view about "like fights like" is very widely held among soldiers, defence scientists, and academic experts on defence.[2]

Then again, there's the phenomenon known as "tank terror", which Soviet minor tactics are designed to exploit.[3] My first moment of action was when I was marker tank in the Tobruk breakout and a very brave German jumped on my back flaps armed with a Molotov cocktail and a crowbar. We had a doubly lucky escape. My troop corporal was covering but his gunner pressed the wrong button and a two-pounder shot went straight through the German and nicked the armour by the small of my back. And the tank being a diesel one, we managed to get the engine fire out. I must admit that ever since then I've suffered a certain amount of "infantry terror".

Seriously, there's a great deal the infantryman on his feet, or better perhaps on his stomach, can do about tanks. And there are increasingly good reasons why he needs to do this. I want to talk about the weaponry, the minor tactics, and finally a defence concept partly or even wholly based on men against tanks.

But first, a word about the reason why. As I previously mentioned, the space between the Inner German Frontier and the Rhine is no longer big enough, particularly deep enough to fight a major battle. It is widely accepted that the Soviet attack, now not so much a steam roller as a steam hammer is going to punch through any normal positional defence somewhere. And there may not be time or space to contain the breakout. The various aggressive defences, such as the "airland battle" or the operational counter-stroke that is being more and more talked about, even in Britain, are desperately expensive. What is more, like the media in a counter-insurgency situation, a mobile defence acts like a sounding-board. It resonates with the offense, thus amplifying the dimensions of the battle and accelerating its tempo -- just the opposite of what is needed.

So, as you have heard, "fortifications" has become a printable word again. And more and more people are looking at the part quasi-guerrilla tactics can play in the defence.[4] And against an armoured offensive, quasi- guerrilla tactics mean "men against tanks".

If I was an infantryman shopping around to equip myself to fight tanks, I'd start by going to London and getting some rigid limpet assault mines, preferably modified for land use. These have to be stuck on the armour but this may not be as difficult as it sounds. Next, I'd go to Washington and get myself some scatterable track-cutting minelets, suitable for hand scattering. These are not lethal but immobilising. Being stick-shaped, they are easier to carry, handle, and lay than the lethal but complex hollow charge minelets. Then I'd go to Moscow and really set about things. First I'd buy some anti-tank grenades. These are stick grenades that you lob high out to about ten metres. They have a hollow charge and a drogue parachute to get them pointing the right way as they come down.

Still in Moscow, I would obtain a RPG7V, which comes in a folding version for airborne troops. In fact, it's now obsolescent, but I don't know details of the successor. With proper training, and this is very important, you have a reasonable chance of a hit on a hull-up tank out to five hundred metres. As every self-respecting terrorist the world over knows, the hollow charge round has excellent multi-purpose effect too. The thing is it's a really handy weapon.

But for tank hunting it needs a very simple fougasse kit, which one could easily carry in a combat jacket pocket. One of those survival-pack fishing kits might even do. For this you lash the launcher to a fence or hedge, or dig it into a bank, pointing across the road or whatever at about turret-ring level. Then you tie a string to the trigger and, with the help of a couple of small pulleys, run it about half a tank's length in the direction the enemy will come; and then just below track height across the road. A lot of people rigged these kits for their bazookas in Second World War.[5]

Moving up the market a bit, you could have Milan if you liked, but from this point of view I'd stay in Moscow and opt for Sagger. A team of two can carry a launch kit and up to six missiles, and the range is now three thousand metres. Anti-tank weapons such as they provide a simple, cheap, rugged armoury of weapons you really can

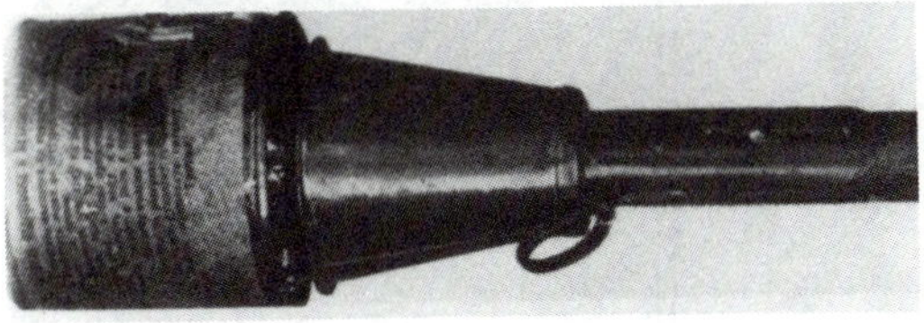

British Rigid Limpet Assault Mine and
Soviet RKG 3 Antitank Grenade
Fig. 9-1

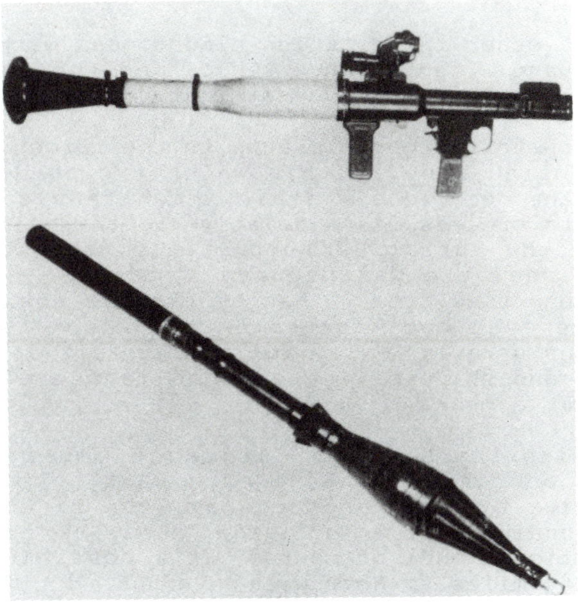

Soviet RPG7V:Launcher and Round
Fig. 9-2

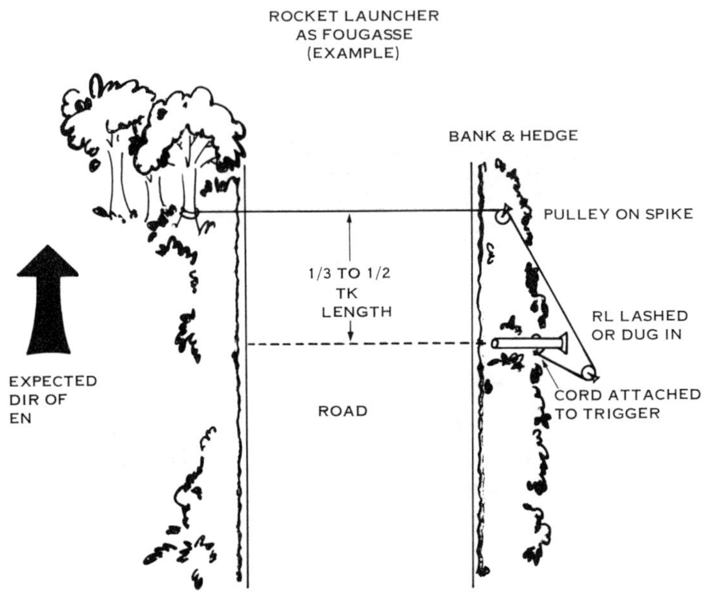

Rocket Launcher Booby-Trap
Fig. 9-3

hump around. One other thing, which almost goes without saying, each small team must have a short-range radio, and the gunner must be on this net. Now let's look at how you might use them.

Once again, let's start right close up to the target. Fig. 9-4 gives an idea of the blind zone, within which a closed-down tank crew cannot see, and the dead zone, within which they cannot fire. The blind zone is more or less fixed relative to the hull; the dead zone goes round with the turret. Unfortunately, the tank I picked to construct this schematic was Leopard 1, which -- at least on the drawings I was measuring from -- has by far the smallest blind zone of any tank I've ever come across. It's always short under the driver's nose, but usually bulges out almost to the dead zone at the sides. The size and shape of these zones is one of the things you need to know about a tank.

Now let us examine the sort of distance you might get with a concealed position behind a hedge or a bank. If you want to hit first time and live to fight another day, the sides are a bad bet. The tracks and running gear -- or large wheels -- give a surprising amount of protection. And, of course, it's possible to miss a tank crossing at twenty miles an hour with an RPG7 at fifty metres. The hull rear is a plum target because it's large and thin. A root up the backside is good against APCs and will immobilize a tank. But it may not put the tank crew out straight away and they may turn nasty.

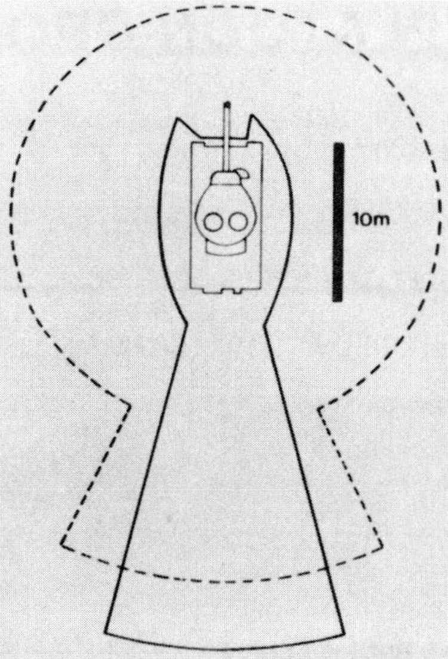

All tanks have a *blind zone* (full line) within which nobody can see the ground, and a *dead zone* (broken line) within which fire cannot be brought to bear. The blind zone rotates with the turret

Tank Blind and Dead Zones
Fig. 9-4

So I think, though you may disagree, that the best angle from which to take a tank is the front. The classical V-front doesn't offer too good a target, but the effort partly to improve swimming, and partly to get good protection on the upper hull with a near-horizontal glacis- plate, has led designers to nudge the tip of the V upwards. The lower front, which is relatively thin, thus becomes more nearly vertical, larger and an excellent target. The ideal way is to take the vehicle coming over a crest.[6] In any event, you need a concealed position just to the side of a narrow defile, preferably covered against prophylactic machine gun fire, and low down.

At longer ranges, both with the rocket launcher and an ATGW, more conventional tactical deployments are the rule. To make a truly formidable tank-hunting organization, however, I would take a normal infantry rifle platoon and add to it an artillery observation party (OP), on their feet of course, and a detachment of half a dozen sappers, each headed by a sergeant. This would provide a platoon of some fifty all ranks, the boosted strength having two purposes: the main one to give the sections a high enough battle strength to mount three patrols of three; and the secondary one to give the headquarters enough men to detach a protection party for the OP or some of the sappers without crippling itself. Incidentally, in this regard, the gunners and sappers would have to let NCOs do jobs normally reserved for officers, such as controlling fire and laying and firing demolitions. They just don't have enough officers to go round and if they did the last thing the platoon commander wants around his neck is a couple of specialists who out rank him.

LIGHT PLATOON GROUP

1 + 1 (SGT) + 43 + (ARTY) 1 S/SGT + 2 + (ENGR) 1 SGT + 6 = 55 (LESS 11 LOB)

FORMS

1 OP PARTY + ESCORT (3 FROM PL HQ)

1 AMBUSH GROUP - PL HQ (-) + 1 SECT + (ENGR) 4

6 TK HUNTING PTLS - EACH 3

ROLES

1. **TANK HUNTING**
2. **CALLING DOWN FIRE/AIR ATTACK**
3. **BLOCKING, MINING, AND BOOBY TRAPPING**
4. **AMBUSH, FIRE SP, MOBILITY DENIAL**

Light Platoon Group Structure and Roles
Fig. 9-5

As for deploying this platoon, I would suggest the following as typical. Two sections would form three tank-hunting parties each -- the section commander's party armed with Sagger, the others with RPG7s (with the fougasse kit). Each section would take one sapper, with a selection of nastinesses; the section commander would decide on the ground which patrol to put the sapper with. The gunner, of course, would tend to gravitate to the kind of place gunners like to go to, and would probably require an escort. The remainder of the platoon -- the headquarters, one section, and three or four sappers including the senior one -- could be tasked to occupy a communications node or defile, covering it with an ambush. Let me stress here that there are a thousand and one ways of deploying such a platoon group but this might serve as a useful start point.

I would additionally suggest that even in good tank country almost every likely enemy thrust line has some patches of close country or bad going in it. These may include broken country with short fields of view and narrow defiles, marshes with flood-banks or causeways, minor water obstacles, defiles with villages in them, or substantial tracts of forest land. For purposes of illustration, therefore, let us take something rather geometrical (Fig. 9-6), a forest area managed in the usual way, with a grid of small temporary tracks or rides; slightly wider and rather straight permanent rides; and a coarser-meshed network of trails, tending to contour, and thus curving and not too related to the pattern of rides.

NET PLATOON AREA (FOREST)

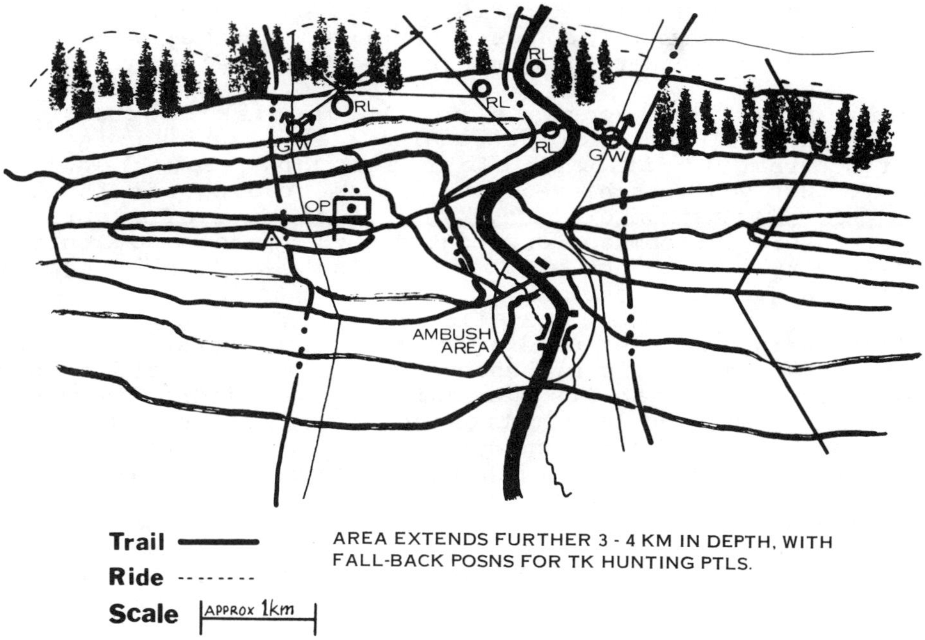

Light Platoon Group Deployment
Fig. 9-6

Although much will depend on the actual ground, of course, I believe -- partly from common sense and partly from historical analogies -- that an element of some kind will be required about every two map squares -- two square kilometres. On this basis, a platoon could have a frontage of about two kilometres and a depth of around seven. This would deploy a company roughly in a square and would mean that a battalion could, as a very rough rule of thumb, cover a twenty kilometre sector. There seems little point in having a tactical reserve in this setting.

To return specifically to platoon deployment, because this is essentially a platoon battle, each platoon area would doubtless contain some kind of defile or node -- the key to passage through the area. This would be the ambush site and focus of the major sapper effort; and it would have priority in the defensive fire plan. If necessary, engineer stores, including anti-tank mines, could be lifted in by helicopter. Once again, the layout of the ambush is completely dependent on the ground; but the aim might be to take out a motor rifle platoon.

This is perhaps a good point at which to stress the ease and value of blocking narrow routes with knocked-out armoured fighting vehicles (AFVs). Theoretically, when, say, a tank blocks a route, another tank can push, or if necessary tow it out of the way. One of the spectacular incidents in film battles is always a tank being pushed over a bridge or an embankment. But in a narrow forest ride or village street, it's not so funny. With a few anti-personnel mines and booby-traps around it, and observed fire on call, a knocked-out tank is a blocked route. I know that tanks can push down trees and buildings, but pushing, or even pulling, another AFV over the resulting mess is another thing again.

So forcing a passage becomes a very high risk operation. The opposition may well decide to go somewhere else, or to bring up mass infantry, dismount it, and clear through. You then hastily disappear. But you've achieved your aim of trading off the minimum of space for the maximum of time. And these kind of pinpricks are just what the Soviet command and control system does not thrive on, there is a very good chance of producing a breakdown of command at battalion headquarters level.

With all this in mind, let us examine a section operation. Remember that the rocket launcher's hit chance begins to fall off beyond two hundred metres and that its maximum is five hundred metres. So anything from four hundred metres upwards becomes a useful ATGM range. Even in very close country, one should be able to get these kind of ranges and better, either along straight rides or from positions overwatching forward slopes. It's very tempting to have initial ATGW positions on the edge of a forest or the bound beyond a defile. But these are likely enemy artillery targets so the game may not be worth the candle. It's obviously important to choose in advance where you'll take the target -- it must be where a single vehicle will block the route -- and to go forward and mess it up before the enemy arrives.

One possible task for the rocket launcher patrols might be to cover the ATGW patrol's target area from a flank. If you can get two knocked-out vehicles one behind the other, the opposition has a problem. And a burst of automatic fire on anybody who gets out doesn't make life easier for him.

Another good bet for a rocket launcher patrol, especially if they have a sapper with them, is to take on a minor ride, or the kind of twisty lane with high thick hedges we have in Britain, where ranges are too short for a guided weapon. The principles are the same, but with the patrol much closer it would help a lot to stop the first vehicle without disclosing the patrol's position. This is where a mix of lightly buried scatterable track-cutting minelets, anti-personnel mines and booby-traps comes in. With luck, particularly given Soviet minor tactics, a second vehicle will close up. If one waits until somebody's dismounted and then opens up simultaneously with rocket launcher and automatic rifles, the desired effect will be achieved.

A word about artillery and air support might be appropriate at this point. The gunner must be responsible for calling down fire and for calling in rotary and fixed wing air support. But the call has to come in the first place from the patrols; and they will have to give target indication for air attack, possibly with a coloured smoke round from the rocket launcher. So some very simple technique -- much simpler than target grid -- is called for. For instance, the FOO -- or FONCO -- might issue each patrol leader with a few target codenames. Having chosen his attack position, the patrol leader could radio a codename and grid reference, plus the grid reference of his own position, to the OP party. The security risk of using short-range radio is, I suggest, less than that of having people charging about all over the forest liaising madly. Then fire would be called down simply by the codename. For shifting or correcting fire, all that would be needed is the codename, a direction (plus, minus, left, right) and a distance.

The alternative tactic, particularly at night, is stalking stationary AFVs. For one's own survival as much as anything else, one must hit first time, and this means going in fairly close with a rocket launcher. Depending, as always, on the ground, in particular on whether there is a covered approach that will take one within the blind zone, it may be better to cover one man forward and use an anti-tank grenade or a limpet mine.

While I do not wish to escalate this talk to higher levels, there is the small matter of insertion and the not so small one of extrication. In this "hammer, anvil, and net" defensive concept, which is one way a switch to quasi-guerrilla tactics might develop, one would have a permanent frontier obstacle zone; a tight positional defence, the anvil, holding a major defile on strong ground; and an armoured or mechanized brigade, the hammer, in depth, with the primary role of delivering a counter-stroke in front of the anvil, and the secondary one of counter-penetration. The net, in this example formed by a light infantry brigade, is deployed at about a tenth of the density of the anvil, with (in this instance) two layers in front, one to either flank, and a fall-back position for the forward battalion. From that rear position it could support the mechanized brigade in the counter- penetration role, or cover the division out and provide the forward net for the next layer of defence.

Since it is fairly essential for the net to be in position before the opposition arrives, insertion should present no particular problem. It may even be possible to leave a vehicle forward until a platoon group has done its reconnaissance and decided exactly what weapons and explosive stores it wants. Alternatively, stores and personnel, if need be, can be inserted by helicopter. It is very tempting to leave, say, a light APC in for extrication of at least the ambush party. All I can say is that I have studied this option in depth and discussed it at length, and it looks as if these vehicles would make the whole group a hostage to fortune. It may be possible to run a vehicle into an agreed rendezvous (RV) or to put a transport helicopter in but I, and others I've discussed this with, believe the best and safest way out is on your feet, with patrols -- or at any rate sections - moving independently until they reach a safe area. It would be extremely inconvenient to have the forward net falling back onto the anvil position just as the opposition was arriving there. So there is a need to deploy the net in such a way that it becomes a network, and each element has another net area to pull back through only a few hours' march away. Some may disagree, but I'm sure the key to the success of any guerrilla-like concept is to avoid dependence on hardware and on conventional ways of doing things. They make you predictable and therefore attackable.

I'd like to close with a word about the training aspect. Evidently, working like this would call for every soldier to have an extremely high standard of weapon training, with the ability to use several weapons. The same goes for fieldcraft and map-reading. At least a

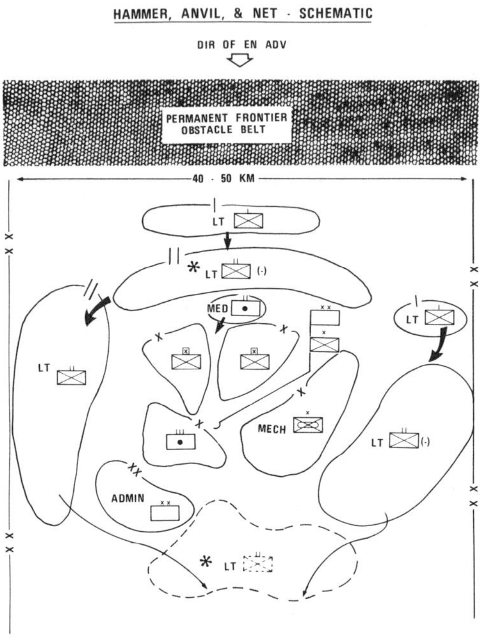

Hammer, Anvil, and Net Schematic
Fig. 9-7

proportion of infantrymen would have to learn elementary sapper skills, so that the very limited sapper effort available could be used on the key position. As I indicated above, patrol leaders would need to be taught some very simple way of calling down and adjusting fire, and of calling for air support. Rather more difficult perhaps, at least three men in each section would have to be of section commander calibre and trained up to patrol leader standard.

But with today's soldier and today's training aids and expertise, none of that should be too much of a problem. The question, which I will simply pose and leave for thought is whether the infantry soldier, who is essentially trained to fight in relatively large groups and at high densities, can retain these essential skills <u>and</u> learn to adopt a whole new way of fighting. I'm quite certain that this or any other infantry battalion could bring at least a company's worth to what amounts to special forces standard, probably given time the whole battalion. What I'm <u>not</u> so sure about, however, is whether a unit can achieve and sustain the flexibility of outlook and breadth of training needed to operate equally well in either of two different ways -- in the anvil or in the net -- as required.

<u>EDITORS' NOTES</u>

1 Richard Simpkin, Tank Warfare: <u>An Analysis of Soviet and NATO Tank Philosophy</u> (London: Brassey's, 1979), 16, 18.

2 Brigadier Simpkin's counter to this is most pertinent and worth some elaboration. In his words: "I have to confess to some puzzlement and intellectual shock when I hear distinguished soldiers and eminent analysts joining forces to preach the doctrine of 'like fights like'. This thinking looks to stem on the military side from the desire for dedicated weapon systems and the trading off of organizational versatility for efficiency in a single role. Along with these go easements of the training problem, especially for conscripts, by focusing on a narrowly prescribed role, and a clear-cut, simple doctrine on how that role should be executed.... Yet carried to extremes this doctrine reduces the land battle to a group of uncoordinated and unrelated exchanges. Tanks fight tanks, infantry somehow closes and fights infantry, artillery is confined to a counter-battery role, fixed-wing aircraft concern themselves exclusively with the air, and helicopters engage in a duel of a kind not so far seen.... From a glance at the major technological innovations in air, land and sea warfare I incline to the view that 'like v like' is an *accident of the state of the art*.... In sum 'like v like' represents a useful constraint on thinking but a sterile and dangerous doctrine. The need is rather *to concentrate all available effort, whatever its nature, in time and space against the opposing element which is critical at that point in time and space*. Simpkin, *Antitank*, 226-229.

3 On the Eastern Front in the Second World War "tank terror" or "tankophobia" was recognized by armies on both sides as a serious psychological disease that had to be overcome through training, often severe and desperate in nature. See Major-General G. Biryukov and Colonel G. Melnikov, *Antitank Warfare*, trans. David Myshne (Moscow: Progress, 1972), 94, 135-136.

4 See, for example, the argument for "adjunct light infantry" in Steven L. Canby, *Classic Light Infantry and New Technology* (December, 1983) Director Defense Advanced Research Project Agency, Contract No. MDA 903-81-C-0207.

5 The modern French Army continues to teach such techniques, and others, in its Combat Rapproché Anti-Char (CRAC) courses. Colonel Andre L. Rilhac, "Commando Training for French Armor", *Armor* (September - October, 1983), 18-20; and Lieutenant-Colonel C. Guizonnier, *Anti-Tank Training in the French Army* (Unpublished).

6 Reverse slope positions will likely continue to afford infantrymen such vulnerable armoured targets.

CHAPTER TEN. INDICATIONS FROM WAR GAMES

METHODOLOGY

by
Lieutenant-Colonel W. D. McKay, CD
Directorate of Land Operational Research (DLOR)
National Defence Headquarters, Ottawa, Ontario

DLOR deals with army operational research and is a part of the Operational Research and Analysis Establishment (ORAE). DLOR is the largest of the directorates in ORAE with about a quarter of the manpower of the establishment; it consists of 11 defence scientists, 11 officers, four NCOs, and four computer support personnel. I am the head of the War Game Section and am, with Dr. Armstrong who heads the Methodology Section, responsible for our wargaming activity. DLOR has two other sections, one of defence scientists and the other a computer section which does most of its work supporting the war game.

War gaming is probably our most important operational research tool; it would also be no exaggeration to say that DLOR is principally an organization that conducts war games. Over the years, we have evolved a generalized and flexible set of war gaming rules and procedures that cover most of the non-moral effects of war. Each set of rules exists in two versions, one that is as detailed as we can make it for research games and one that is less detailed but much faster for training games.

The research game is a manual game with a great deal of computer assistance. Despite the impression given by the amount of computer equipment, it is emphatically not a computer game. By this I mean that it is not a game which is started and left to run by itself until it comes to an answer. All important decisions are made by humans; specifically, experienced senior army officers from a mix of combat and support arms. In short, we believe that the computers are doing what they do best -- they carry out the assessments and record the endless quantities of data which the games produce. The humans make the sort of decisions that cannot be pre-programmed -- for example tactical decisions -- and the controller controls the flow of the game making sure that the interactions are all within the spectrum of the militarily possible.

From January, 1979, until March, 1982 we were involved in a research war game series called Bronze Talon. These war games were carried out in order to support Canadian army combat development, involving equipment, tactics, concepts, and organizations. The intent of the series of four games was to play a Canadian brigade group:

a. In defence (Games I and III) against a Warsaw Pact second echelon motor rifle division; and then,

b. In offence, a counter-attack situation (Games II and IV), against a Warsaw Pact first echelon motor rifle regiment (minus).

Games I and II played the current mechanized brigade group in order to measure the effectiveness of a new brigade organization termed the System Study Formation. The new brigade was then played in the two following games.

Our current series, nicknamed Iron Viper, is an examination of the System Study infantry battalion fighting as part of a corps. What we are especially interested in is the evidence we obtain about the brigade, division, and corps support that this battalion would receive. What we have elected to do, therefore, is play a series of games in which the battalion, with different amounts and kinds of support, defends against the enemy force that we feel it can stop. After a good deal of thought, we decided to ignore quite a lot of weapons effects -- we are not playing helicopters, fixed wing air, target acquisition (by radar or airborne platforms), air defence, resupply, communications, or chemicals. We feel that we can safely ignore these areas because it is a comparative series and we will be comparing like with like; in fact, it is rather like a laboratory experiment, with similar strengths and weaknesses.

In the DLOR research war game, we have attempted to include every facet of land-air warfare that we can measure. This desire to include everything has been fed by the availability of computer support, which, in turn, has expanded to satisfy the new desires (a process familiar, I'm sure, to all of you who have worked with computers). At the present moment we can claim to have rules for every aspect of the land-air battle except for moral factors. If it is true, as Napoleon is reputed to have said, that "war is three-quarters moral factors," then our war games deal with part of the remaining quarter. As we have no conception (and neither, I suspect, does anyone else) of how to war game morale, leadership, training and those other factors that actually do win or lose wars, we are not prepared to add these factors to our games. Any attempt to do so would probably be just wishful thinking.

How good are the rules? Well, we try to make the best and most accurate rules we can but we really don't know how accurate they are (and again, I suspect, neither does anyone else). We can thus, in truth, only claim to be operating somewhere in Napoleon's remaining 25 percent and not covering it all by any means. War gaming is, after all, a rather uncertain business and war game results must be examined in the light of experiments, field exercises, military judgement, and the historical record.

I would like to point out that the two series of DLOR research games that will be used as examples by the following presentations have been most valuable in terms of giving us better insight as to why certain equipment and their tactical use might be more successful on the battlefield than other equipment.

THE ARTILLERY THREAT

by
Major G. H. Sawtsky, CD
Directorate of Land Operational Research (DLOR)
National Defence Headquarters, Ottawa, Ontario

If one looks at the organizational diagrams of both a Soviet motor rifle division and a tank division, one can see that our potential adversary possesses a large number of tanks and infantry fighting vehicles. Let us, however, look a little closer at the artillery resources organic, and I stress the word organic, to both of these divisions. Note, for example, that each motor rifle regiment has a battalion of eighteen 122mm SP or towed artillery pieces, and three 120mm mortar batteries at its disposal; the tank regiment has eighteen 122mm SP as its organic artillery. Add to these organic resources the divisional artillery regiment, the multiple rocket launcher battalion and the free flight rocket (FFR) battalion, and you have the grand total as shown in Fig. 10-1.

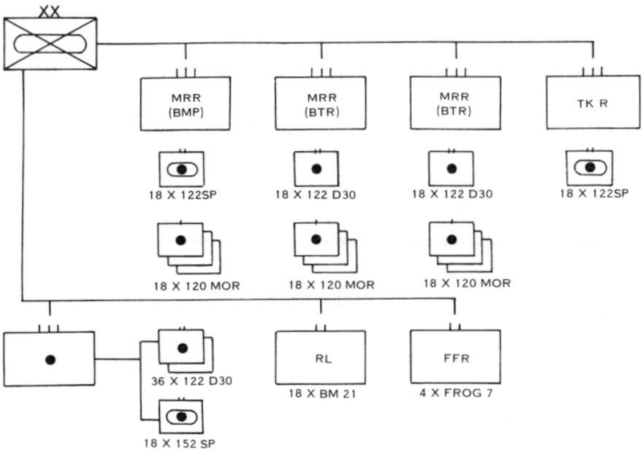

120 mm Mortar	122 mm M1974 SP	122 mm D 30	152 mm M1973 SP	BM 21 MRL	FROG 7* FFR
54	36	72	18	18	4

* BEING REPLACED BY SS-21

Motor Rifle Division Artillery Organization
Fig. 10-1

A look at the tank division shows very similar totals except for an increase in 122mm SPs. Putting all these facts together we see that, yes, the enemy has a lot of tanks and infantry fighting vehicles, but he also has a lot of artillery organic to his motor rifle and tank divisions. If the 202 artillery pieces of a motor rifle division (which do not include 82mm mortars) concentrated their fire, they could deliver in one volley 66 tons of steel.

I would now like to take a brief historical glance at Soviet artillery. According to Chris Donnelly in a recent article in International Defense Review,[1] Soviet military doctrine since 1945 has sought to reduce the battlefield to a series of comprehensive statistical calculations to enable Soviet generals to plan any future campaign. As a result of meticulous and detailed research into their experiences between 1941 and 1945, their experiences on exercises since then, and the experience of combatants in so-called modern local wars, the Russians have established fixed "norms" governing the conduct of tactics and operations. These norms are particularly applied to artillery and the artillery regulations manual is to a large extent based on these norms.

Furthermore, the discipline of a military doctrine, and the constant research which the existence of that doctrine inspires, tends to make the Russians much more precise in the use of terms than Western gunners. Soviet gunnery is more rigidly structured and more dependent on theoretical calculations. Established norms, as indicated in the following figure, determine the number of artillery or mortar shells of different calibre that should be used per hectare on selected target areas such as dug-in infantry or battery positions. Degrees of neutralization, suppression, or destruction are explained at the bottom of Fig. 10-2.

	Guns and howitzers				Mortars			MBRLs	
Calibre (mm)	122	130	152	203	120	160	240	up to 140 mm	over 140 mm
Carefully prepared defence	200	200	150	60	200	100	50	320	100
Hasty defense	150	150	110	45	140	85	45	240	80

Number of shells of different calibres per hectare of target area that must be fired against dug-in positions and strong points, so as to achieve suppression. These norms are valid at ranges up to 10 km.

Range (km)	6	8	10	12	14	16
Weapons						
155mm howitzers	143	179	229	317	-	-
152mm gun howitzers	142	168	214	283	356	450
130mm field gun	178	209	270	332	408	488

The number of shells of different calibres to be fired at various ranges to effect the suppression of a battery of towed guns.

Range (km)	4	6	8	10	12	14	16
Weapons							
152mm howitzers	196	241	334	438	645	-	-
152mm gun-howitzers	156	245	312	407	547	698	902
130mm field gun	109	187	294	458	655	808	962

The number of shells of different calibres to be fired at various ranges to effect the suppression of a battery of armoured SP guns.

	Rifled weapons			Mortars		Rocket Launchers	
Calibre (mm)	122	130	152	160	240	Medium	Heavy
No of shells needed	260	240	200	140	110	270	200

The number of shells needed, at all ranges up to 10 km, to achieve the destruction of a NATO Lance Launcher

- **Neutralization** (neitralizatsiys) - minimal damage but paralysis of the defense during and for a short time after the bombardment.
- **Suppression** (poddavlenie) - 25% destruction of men and equipment, seriously reducing the effectiveness of the target.
- **Total destruction** (unichtozhenie) - 60% destruction or more of men and equipment - the target ceases to be capable of military activity.

Soviet Artillery Norms
Fig. 10-2

Current Soviet doctrine divides artillery support for an offence into three distinct phases:

a. Preparation fire, which is the fire delivered before the offensive starts so as to soften up the defence;

b. Close support, defined as fire delivered during the assault; and,

c. Accompaniment, the fire delivered by guns accompanying the attacking troops into the depth of the defence.

Let us now have a quick look at preparation fire in the Iron Viper war game. During the "nominal" force game, Red (representing Soviet forces) concentrated 164 artillery pieces and four 82mm mortars. During the 30 minute fire plan these tubes fired 408 tons of ammunition at the Blue defender (representing Canadian forces), in this case a dug-in infantry battalion. The battalion lost 102 infantrymen as a result. During the "breakthrough" game, Red concentrated 342 artillery pieces against the same Blue defender, softening them up with 936 tons of steel. Over 200 soldiers were lost. One must bear in mind, however, that 342 artillery pieces on a four kilometre breakthrough front only equates to about 86 guns per kilometre; somewhat short of the 100 tubes per kilometre the Red Army normally tries to achieve. Fig. 10-3 shows some of the historical examples of Soviet concentrations during the Great Patriotic War.[2]

TACTICAL DENSITIES IN SOVIET RIFLE CORPS IN WW II OFFENSIVE OPERATIONS

OPERATION & DATE	CORPS	ZONE OF ATTACK (km)	SECTOR BREAK-THROUGH (km)	DENSITY PER KM OF SECTOR OF BREAKTHROUGH		
				RIFLE BN	GUNS & MORTARS	TANKS & SELF PROPELLED ARTY
KURSK COUNTEROFFENSIVE JULY 1943	XXXVI	3	3	9	128	25
	XVI	22	7	3.9	156	25
	VIII	4	4	6.7	221	16
BYELORUSSIA JUNE 1944	XXXVI	5.5	5.5	4.9	138	13
	XXXV	5	4	4.5	139	32
	XVIII	7.5	5	3.6	180	17
LVOV-SANDOMIERZ JULY 1944	XV	4	4	6.2	240	3.7
VISTULA-ODER JANUARY 1945	LXXXIX	4.5	4.5	6.0	148	13
	XXIX	3.5	3.5	5.1	272	23
BERLIN APRIL 1945	LXXX	5.0	2.5	4.8	342	25
	IV	2.5	2.5	7.2	251	44

From 76mm caliber and higher

MASSING OF MEN AND WEAPONS IN THE SECTORS OF THE MAIN EFFORT OF SOVIET ARMY OPERATIONS DURING WW II

OPERATION & DATE	ARMY	FRONT km	BREAKTHROUGH SECTOR km	DENSITY PER KM OF SECTOR OF MAIN EFFORT			
				PER RIFLE DIVISION km	ARTILLERY (From 76mm & up)	TANK & SP ARTY DIS	TANK & SP ARTY TOTAL
KLIN DEC 1941	THIRTIETH	80	--	7	11	1	1
KHARKOV MAY 1942	SIXTH	80	18	3	37	7	25
BOLKHOV JULY 1972	SIXTY-FIRST	80	10	2	91	11	33
COUNTEROFFENSIVE AT STALINGRAD	FIFTH TANK	35	14	3	72	12	27
BELGOROD-KHARKOV AUG 1943	FIFTY-THIRD	7	7	1	161	11	41
ORSHA JUNE 1944	ELEVENTH GUARDS	35	10.5	1.1	221	22	46
VISTULA-ODER JAN 1945	SIXTY-NINTH	55	7	0.9	282	27	73
BERLIN APRIL 1945	EIGHTH GUARDS	13	7	1.0	320	29	83

Soviet Historical Force Concentrations
Fig. 10-3

Red forces are not just content with simply establishing quantitative superiority in guns or tanks or in massive fire for the preparation phase; they also use large quantities of smoke through all phases of the offensive to ensure that their movements are covered. Our observers consequently cannot bring down adjusted fire upon assaulting troops and our anti-tank weapons are at the same time screened or rendered ineffective (I believe that the majority of us here are quite aware of what smoke can do to TOW fire). Throughout our games we saw that Red has sufficient smoke munitions to screen off any area he chooses. His units of fire also allow him to maintain a continuous smoke screen for long periods of time.

Unfortunately, in countering the overall enemy threat, we tend to concentrate on tanks, often paying less attention to what I consider the largest threat of all: namely, the guns. We must find ways of suppressing enemy artillery while enhancing the capabilities of our own guns. Red with his massive amount of artillery can breach holes in our defences and at the same time neutralize our artillery. While countering enemy guns has traditionally been considered a purely artillery matter, I do not believe this will suffice today. It must be seen as a common command or operations problem. What is needed is a deliberate and well-coordinated counter-battery (CB) plan involving not only artillery staff, but also intelligence staff, electronic warfare staff, and air staff.

The first requirement, of course, is to locate Red artillery to an accuracy which will permit engagements with a reasonable chance of success. This presupposes that our land forces must have a full range of target acquisition devices, such as counter-gun and counter-mortar radars, remotely piloted vehicles, sound ranging, infra-red flash spotting and any other means that will help locate Red artillery. Electronic warfare resources must also play a vital role in degrading the effectiveness of Red artillery. These resources must concentrate on disrupting Red target acquisition batteries by jamming their radars and reporting links. Similarly, Red regimental artillery group and divisional artillery group fire coordination links should be jammed, as well as all forward observer communications. As part of such a coordinated counter-battery plan, ground attack aircraft should additionally be targeted against Red artillery rather than tank concentrations as is often now the case.

As for our own guns, we require sufficient and suitable weapons systems with a range to deal effectively with Red artillery. Practices such as "shoot and scoot", and increased dispersion without loss of the ability to concentrate fire must become routine to ensure survivability. Armoured protection will also be vital, as with the advent of improved conventional munitions (ICM) digging-in guns may be a waste of time. The tendency to fire Blue guns on an almost continuous basis, of course, raises the further question of the ability of artillery units to deliver fire on a sustained basis, 24 hours a day. Though benefits of automation suggest that a single delivery system could be manned, maintained and secured by a crew of two to four soldiers, it is unlikely that this small crew could remain effective for more than 48 hours.

It might now be worth stating that recent research war game results from the Bronze Talon series underline the fact that artillery on either side remains the greatest killer on the battlefield. That artillery should account for the largest percentage of kills on the battlefield is not a startling new revelation. The history of 20th Century warfare bears this out, as does data from our other games. The most effective weapons system on the battlefield, in fact, appears to be artillery.

ON INFANTRY DEPLOYMENT

by
Major G. K. Jensen, CD
Directorate of Land Operational Research (DLOR)
National Defence Headquarters, Ottawa, Ontario

In this section I shall cover APC-borne and MICV infantry tactical deployments. The second part of my presentation will highlight certain problem areas and tactical questions related to infantry anti-armour weapons employment. Most of what will be presented is based on data collected during the Bronze Talon and the Iron Viper series of research war games, supported by what one might call judgements and insights contributed by various players and the controllers during game play and post game analysis.

Let us first look at the types of vehicles that were used. In Bronze Talon I and II, the Blue force was equipped with our standard M113 APC, which carried 10 infantrymen and was armed with a .50 calibre heavy machine gun (HMG). In addition, Blue was also equipped with the Lynx command and reconnaissance vehicle, a varient of the M113, which carries a crew of three and has an electrically fired HMG. In Bronze Talon III and IV, as well as in the Iron Viper series, the foregoing tracked vehicles were replaced by the Canadian infantry combat vehicle (CICV). This imaginary vehicle is, in many ways, similar to the American BIFV in terms of mobility and protection. It carries a crew of 10 infantrymen including the driver, the gunner, and the crew commander. It is armed with a 25mm cannon, with thermal imaging sight, that has an effective range of 1000 metres against the BMP and a maximum range of 2500 metres. Unlike the BIFV, the CICV does not have an integral anti-tank guided weapon.

Having briefly described the vehicles, it is appropriate to mention which organizations were equipped with each. In Bronze Talon I and II, the M113 was the standard vehicle found in infantry rifle companies. The Lynx was found in the infantry battalion reconnaissance platoon, the brigade reconnaissance squadron as well as in the reconnaissance troop of the armoured regiment. As previously mentioned, in Bronze Talon III and IV and the Iron Viper series, the CICV replaced both M113s and Lynx.

Before discussing tactical concepts it might now be worthwhile to quickly review the main uses of the APC and the CICV. An APC such as the M113 is primarily a taxi which gives troops the capability to move about the battlefield in some degree of protection from small arms fire and shell fragments. The APC also possesses some limited firepower in its HMG or GPMG. The APC was not designed for troops to fight mounted from it, although it is possible to do this in very special circumstances. The CICV, on the other hand, was conceived as a true fighting vehicle that would permit the infantry to fight from the vehicle without dismounting until the last possible moment, if at all. We envisage such a vehicle as having firepower, protection, and mobility in excess of that of the standard taxi APC.

German Marder MICV
Fig. 10-4

Current tactical doctrine for the APC basically dictates that carriers are generally hidden away from a defensive position but reasonably close in a "hide". It is normal practice to dismount the majority of the HMGs or GPMGs and deploy them on the ground, while sending the empty APCs back to the hide where they remain out of battle. This procedure reduces the amount of equipment on a defensive position and gets the vulnerable M113 into safer areas. Sometimes, but not often, APCs are located within a defensive position. As for Lynxs, they are generally forward with reconnaissance units that are allocated to a brigade screen or guard,[3] whichever one a commander decides to employ. We usually find that they are subsequently tasked for rear area security or flank protection, provided they survive and make it back.

There are advantages and disadvantages to siting APCs on or off a defensive position:

a. On a defensive position:

(1) Advantages

(a) crews and vehicles are co-located; and

(b) maximum firepower is available and less time is involved if ordered to redeploy.

(2) Disadvantages

(a) there are more targets in any given defensive position;

(b) the position is probably more difficult to camouflage; and

(c) the mobility of the vehicle is not really utilized as part of the overall defence plan.

b. Off the position:

(1) Advantages

(a) vehicles are generally not subjected to large amounts of artillery fire;

(b) maximum utilization of the combination of mobility and firepower is achieved,

(c) there are fewer targets on the defensive position; and

(d) the position is less difficult to camouflage.

(2) Disadvantages

(a) troops are separated from their vehicles, and

(b) command and control is more difficult.

In the Bronze Talon I series, APCs were generally deployed in hides away from main defensive positions while Lynxs were forward as part of a brigade guard. In Bronze Talon III, substantial numbers of CICVs were deployed on or near defensive positions rather than in hides for reasons I will outline later. It is significant that APCs deployed within defensive positions or near them generally became casualties, the majority of which were caused not by direct fire weapons but by artillery. Those which survived artillery fire contributed little as they were so suppressed they could not move off the position; nor could their crews function to full capacity. This should not be surprising, however, in view of what my artillery colleague previously outlined. Interestingly, it was also found in both Bronze Talon I and III that as soon as Blue direct fire opened on Red from a previously undetected position, the latter immediately responded with artillery mixed with smoke, with the same effect already mentioned.

The conclusion appears to be quite evident. If APCs or CICVs are located in main defensive positions they are going to be destroyed quickly or will be neutralized so that they are of little value anyway. It would appear that the best solution is to keep them off the position. The following figures serve to illustrate this point:

a. in Bronze Talon I over half of all Blue APCs destroyed were destroyed by artillery fire;

b. in Bronze Talon III almost half of all Blue CICVs destroyed were destroyed by artillery. Most of the artillery kills were, of course, due to top attack improved conventional munitions (ICMs) and were attained on CICVs deployed on or near battle positions to provide cannon support; and

c. in the Iron Viper series of seven games played to date, on average, nearly one third of all APCs or CICVs were destroyed by artillery.

On a related tactical point, though, anti-armour weapons certainly have a place as part of the weaponry in the guard battle. A commander in allocating them must carefully weigh likely losses against subsequent taskings that he might have for these weapons. To illustrate this point, in Bronze Talon I, the majority of the TOWs allocated to the guard battle were lost and thus were not available for the main defensive battle, even though they were sorely needed. During the guard battle in this game, as well, the M113/Lynx .50 calibre HMG combination was badly outclassed. Lynx HMG, for instance, only destroyed six BMPs and three other types of APCs, which resulted in an unfavourable reciprocal exchange rate. Other members of the M113 family destroyed two BRDMs[4] and a BMP while large numbers of them were destroyed by Red forces. It would thus appear that the HMG is not an effective anti-APC weapon when mounted on APCs and reconnaissance vehicles or deployed on the ground.

In Bronze Talon III and the Iron Viper series, the infantry were not issued a direct fire weapon that could fire rapidly from concealed weapon positions to kill APCs at shorter ranges. Because of this deficiency, Blue tended to deploy CICVs on or near battle positions to add their weight of fire to the defence (again, severe losses were incurred from ICMs). Another reason the CICVs were so necessary to the defence was that they were the only organic infantry weapons system equipped with a thermal imaging sight that could detect and engage the enemy through smoke. It would appear then, that if CICVs are to be deployed well away from defensive positions, the infantry still requires some kind of rapid firing, direct fire weapon with a through-smoke capability to deal with enemy IFVs once they have closed to within small arms range.

If CICVs must be sited well away from defensive positions, how then can they be effectively employed in the defence? In fact, there might be a couple of ways in which they can be used in the mobile fire support role. One, they might be deployed well forward of, and to the flanks of, main defensive positions from where they could cause attrition, disruption and confusion while drawing the enemy onto concealed, dug-in infantry positions (where he becomes involved in a fight that he doesn't want). We must bear in mind, however, the losses that screens and guards often incur. Two, alternatively, and the most likely way we see CICVs being employed, is that they could be deployed from hides to the flanks and between gaps in company positions, engaging the enemy as he comes forward or when he gets in really close. By doing this you conceal defensive positions and do not attract enemy artillery fire onto them. Secondly, you utilize the mobility and firepower of the vehicle as part of the overall plan. It was found, particularly in the Iron Viper series, that CICVs actually in defended positions were usually so supressed or neutralized by artillery that they could seldom fire; and when they did manage to do so they were usually destroyed by the return fire of enemy tanks, BMPs, and ATGM. When they were employed as a mobile fire support element, however, they were particularly effective when shooting and scooting.

If employed as mobile fire support, it is not necessary to concentrate, say, a full company's worth of CICVs in one location. Rather, they could be decentralized into platoon elements, bearing in mind the problem of command and control, still able to hit the enemy from a flank. The important point is that the CICVs themselves do not have to be concentrated, but their fire must be concentrated. Obviously there is a price to pay if we use them in the manner described. It may even be necessary to change rifle company and platoon organizations. The most critical aspect is that one risks the possible loss of CICVs, so absolutely vital to the transport of infantry. Nevertheless, in my opinion, that risk is acceptable in order to gain the advantages offered by the mobility and firepower of the CICV.

As an adjunct to the mobile fire support concept, is the use of the APC or CICV as an anti-helicopter weapon. Both the APC with its HMG and the CICV proved to be effective killers of Red helicopters. In Bronze Talon I, for instance, 35 helicopters were destroyed by HMG fire alone (Red having just happened to fly over concentrations of Blue APCs). In Bronze Talon III, the CICV destroyed 17 helicopters, while HMGs accounted for another 15. If APCs or CICVs are an integral part of a defensive position they may not be as effective against helicopters, however, particularly if they are being subjected to enemy artillery fire. APCs and CICVs not sited in defensive positions give a commander a resource that can be quickly deployed to counter or contain enemy heliborne assaults. If APCs or CICVs are deployed within defensive positions, on the other hand, it will probably be difficult to extract them for use in an anti-heliborne role.

In summary, it is advocated that CICVs are generally best employed in the defence as mobile fire support, thereby taking full advantage of their mobility, protection, and firepower. The dilemma for a commander is that CICVs within defensive positions are subject to heavy artillery fire, especially from ICMs, at a time when he does not really wish to lose the firepower of their 25mm cannon. Also, if he sites CICVs in amongst his infantry, he will probably have to restrict their opening range of fire or risk drawing enemy fire onto his infantry sooner than he would wish. Yet, if he sites his CICVs elsewhere, he risks command and control problems and the possible loss of transport for his infantry.

Turning now to the offence, I must caution you that I do not intend to develop it as fully as the defence. Basically, the APC or CICV in the offence is a taxi to get the infantry into position where he can close with and destroy the enemy. The CICV, it is hoped, will offer more protection, increased mobility and firepower than the M113 family of vehicles. In recounting war game results, I would remind you before I go on that Bronze Talon II and IV each incorporated a brigade level counter-attack, it is also important to note the brigade counter-attacked in very close, hilly country, which certainly affected weapons employments.

In Bronze Talon II, Blue force lost nearly 100 APCs to all causes. It was clear that the M113/HMG combination was outclassed because it destroyed very little in return. In Bronze Talon IV, Blue force lost well over 100 track-type vehicles, a considerable number of which were CICVs; nearly half of all CICV casualties were due to direct fire. What is significant, though, is that while there was a considerable difference in the numbers lost to direct fire, identical numbers were lost to Red mines. Red was permitted in Bronze Talon IV to use artillery and rocket delivered scatterable mines much more freely and in far greater quantity. As a result of this, Blue was continually mired in minefields. Mobility was lost and he became a marvelous, immobile target for Red weapons systems such as the tank, the long range ATGM and helicopters.

Regardless of whether one uses a CICV or an APC, the scatterable mine poses a very real problem in respect of vehicle mobility. The most obvious solution to this threat is to prevent the mine from ever being delivered but that is dependent on the CB effort. Another possible way of ensuring our mobility is to advance up routes that the enemy must keep open himself (the problem here is how to determine which routes these are.) Blue eventually regained his mobility in Bronze Talon IV, but only after throwing almost all of his artillery, including his close support regiment, into the CB battle. Even then, however, Blue's advance could only be sustained under cover of tank and attack helicopter supporting fire. Generally, the CICV could not advance relying on its own firepower alone. It would appear, then, that it is not so much a question of the type of vehicle that one uses as it is the ability to maintain its mobility. This, in turn, is highly dependent on one's ability to suppress enemy artillery.

To provoke further discussion on the CICV and the APC, I would question whether the Canadian infantry really needs a sophisticated vehicle like the CICV when a taxi-type APC, with good mobility and protection, might suffice. Possibly a combination of both may satisfy our requirements. The approach we might adopt is to separate the cannon requirement from the troop-lift requirement. One possibility would be to have in each platoon three APCs of the troop carrying type and perhaps one CICV with a cannon to support movement; alternatively, all company headquarters vehicles could be CICVs (this is not really anything new because a very similar type of organization was used by German panzergrenadier units in the Second World War). Transport APCs could thus remain in hides close enough to move and quickly redeploy foot infantry, but far enough away to avoid destruction by enemy artillery fire. CICVs could at the same time fire in support of defended localities from a distance without fear of certain destruction by artillery. One last thought with regard to CICV cannon-type vehicles is that at battalion level there could be created a CICV cannon platoon, which would give the commander a resource that he could deploy when and where required, particularly to counter the helicopter threat.

I would like now to turn to anti-armour weapons. In Bronze Talon I and II, the Carl Gustav was employed at both platoon and company level while the improved TOW vehicle (ITV) was used in the infantry battalion anti- armour platoon. However, in Bronze Talon III, IV and the Iron Viper series, the Milan medium anti-armour weapon was employed at platoon and company level as well as in the battalion anti-armour platoon. The ITV/TOW was employed as a higher formation weapon whose sub-units were generally allocated to support the brigade.

Tow Firing
Fig. 10-5

None of the foregoing anti-armour weapons in the Bronze Talon series had a capability to engage through smoke; the ITV/TOW alone could detect targets through smoke, but could not guide its missile through smoke to the target it had detected. It quickly became evident that the most effective way to neutralize anti-armour weapons, especially wire guided ones, was to smoke them off. Red employed smoke in large quantities as part of his preparation bombardments and also on a very large scale to cover his movements. Blue anti-armour weapons were consequently often effectively neutralized long before Red was within range. The Blue commander was thus faced with the choice of redeploying some of his major weapon systems before battle had really been joined, or leaving them where they were in their original positions hoping to get a target in any windows through the smoke.

The simplest solution to this problem, it would appear, would be to equip all major anti-armour weapons systems with thermal imaging sights. However, this measure would not entirely solve the problem for wire guided weapons; improvements in the guidance system would also be required to permit the sight to track the missile through smoke. The effectiveness of enemy smoke on the battlefield is illustrated by the fact that in the Bronze Talon series the overall exchange ratio for TOW was quite low indeed. In the Iron Viper series, however, the exchange ratio was more favourable; this was because a tracking system that worked through the smoke and a thermal imaging sight that allowed a TOW gunner to detect and guide the missile was used. Also, whereas the ground in the Bronze Talon series was very close and hilly, in the Iron Viper game it was very open and generally flat; this, of course, optimized the ITV/TOW weapons system. It should be emphasized, too, that the tactical employment of the ITVs in Iron Viper was also changed and improved, which also accounts for TOW's better performance. The fact remains, nonetheless, that the ITV/TOW weapons sytem can be most effective, given that it can see through and guide its missile through smoke. By way of comparison, the Milan in both Bronze Talon and Iron Viper war game series gave a generally disappointing performance.

The Bronze Talon and the Iron Viper series again clearly demonstrated that for any weapons system to be effective, it must be properly sited. In Bronze Talon I and III, ITVs were generally located in company and platoon positions wherein they derived some protection but became subject to three other problems. First, because infantry positions are not normally sited with 3500 metre range fields of fire, the ITVs were usually too close to the enemy when they opened fire. Second, there was the problem of that old nemesis, Red artillery; firing ICMs, it destroyed far more ITVs than did direct fire weapons. If ITV or Milan opened up near their maximum ranges and their fire was detected by Red, he immediately responded with heavy artillery fire on the position. As a result, anti-armour weapons were either destroyed or heavily suppressed (as were the infantry in these positions) long before they could contribute to the battle. Third, anti-armour plans were based on a series of killing zones with long fields of fire. Two interesting effects occurred here. For whatever reason, Blue often prevented Red from getting into the killing zones Blue had set up, thereby negating Blue's fire plan and forcing Red to bypass and go elsewhere, often into areas that were not covered! In addition, Red, not being totally stupid, deliberately chose not to go into large open areas that he concluded would be used as killing zones. In either case, Red often used his artillery on known or suspected Blue positions on the edges of killing zones, with results similar to those previously noted, while he went elsewhere avoiding decisive engagement. Interestingly, at no time in any game was an ITV/TOW weapons system ever directly threatened at short range by a BMP or infantry.

The ITVs in the Iron Viper series were a divisional resource, and, in consequence, were positioned without being tied to any particular company deployment. They were sited to have long range fields of fire into enemy flanks, and, as the battle developed, were redeployed from position to position to counter the threat. The ITVs were thus usually able to fire at their best range while at the same time avoiding the consequences of Red artillery fire. Likewise, they did not attract artillery fire onto infantry positions. By employing these sort of tactics, along with improved technical capability, more favourable exchange ratios were achieved. Blue also sited his Milans of the battalion anti-armour platoon well away from dug-in infantry positions; their performance was not as significantly increased, however, partly for lack of a thermal sight and an appropriate tracking system. Milans, when they did get windows in the smoke, were usually presented with targets that were 300 metres or less away from them; because it is difficult to gain control of the Milan missile under 300 metres, they often could not or did not fire.

At this point, the matter of equipping rifle companies with wire guided weapons, such as Milan, might be examined. The logic of providing rifle companies with Milan, which has an effective range of between 300 to 2000 metres, when other rifle company weapons have a maximum range of 500 to 600 metres is indeed questionable. In Iron Viper, for instance, rifle companies were generally deployed to extremely good effect in reverse slope positions. However, such positions with their short fields of fire, though ideal for small arms, did not maximize the potential of the company Milans. Company and platoon commanders are thus faced with the following dilemmas:

a. optimizing Milan's effectiveness by positioning troops on forward slopes or crests and so increasing the risk of destruction of the bulk of infantry by long range direct fire weapons and adjusted indirect fire;

b. positioning troops on reverse slopes to optimize infantry small arms effectiveness, yet severely limiting Milan's effectiveness; and

c. separating Milan and infantry small arms thereby creating difficulties in both deployment and in command and control.

A further limitation in respect of Milan is related to cover. In fact, its inability to fire from under cover makes its inclusion in the rifle company suspect. Once it is detected, lack of overhead cover makes it easy to suppress; in short, when subjected to artillery fire it can no longer protect the infantry from armoured attack, which is precisely the reason why it is at company level in the first place. Last, there is the problem of minimum engagement range; below 300 metres it is difficult to capture the Milan missile and guide it on target. As the rifle company and platoon battle is normally fought at around 300 metres range, why equip a company or platoon with a weapon that is difficult to use under 300 metres? The evidence, nonetheless, seems to indicate that infantry rifle companies and platoons require a direct fire weapon capable of producing rapid fire within the ranges compatible with other small arms and, hopefully, firing from cover.

ON COMBINED ARMS

by

Dr. Patrick Armstrong
Directorate of Land Operational Research (DLOR)
National Defence Headquarters, Ottawa, Ontario

The subject of my presentation is the fashionable word "synergism". This word, which comes from medicine, is defined in the Concise

Oxford Dictionary as: "The combined effect of drugs, organs, etc, that exceeds the sum of their individual effects." In other words, synergism is a word to be used in cases in which one and one equals three.

All of you know that weapons must be used in combination and in balance. Indeed, this is a military "motherhood" statement. Like most "motherhood" statements, it is also so commonly accepted that it has become part of every soldier's mental furniture with the result that it is not often thought of consciously. By using data generated from the Iron Viper series of research war games, I will demonstrate the importance of synergism so that it becomes indelibly impressed upon your mind.

I will be taking figures from four Iron Viper games in order to illustrate my point. I should take this opportunity to state that all games were played for 140 minutes and, therefore, all numbers you will hear are for exactly the same amount of time. I am going to look at Iron Viper III which employed a System Study infantry battalion with artillery support, Iron Viper VI which employed a System Study infantry battalion with eight tank destroyers and Iron Viper VII which employed a System Study infantry battalion with twelve ITVs. These three games I will call the "non-system games" because these weapons -- artillery, tank destroyers and ITVs -- were used individually and not in combination with each other. From these numbers composite non-system results will be created and they will be compared with the results from Iron Viper VIII in which a System Study infantry battalion had the support of all three -- artillery, eight tank destroyers and twelve ITVs. If there were a synergistic increase resulting from the combined effects of these weapons it should appear in an increase of Iron Viper VIII results over those of the composite results. One of the purposes of the Iron Viper series was, in fact, to get a measurement of this synergistic effect, and that is the reason why the System Study infantry battalion has had different bits and pieces added to or taken away from it.

In comparing the composite non-system results and the system game, two measures of effectiveness will be used: the "cost" to Blue in personnel casualties of killing Red personnel, and the "cost" to Blue in weapons and vehicles of killing Red weapons and vehicles. These are simply different ways of putting the familiar and common measurement of exchange rates. Iron Viper III gives us the individual results attained by a System Study infantry battalion with artillery support. From Iron Viper VI we can get the effects of the tank destroyers and from Iron Viper VII we can get the effects of the ITVs. In all of these games the System Study infantry battalion had the same number of CICVs and Milans. A slight difficulty arises here because it is clear that in Iron Viper III the CICVs and the Milans were the principal anti-armour weapons whereas in Iron Viper VII, for example, these systems would have had to "compete" with the much more effective ITVs for kills. Thus, in Iron Viper III, the Milans and CICVs would have done more damage to Red than they did in the other two non-system games. In order to minimize this effect, I took the effects of CICVs over these three games and averaged them and did the same with the Milan effects. So, in order to create the amalgamated non-system effects, I added together:

a. the kills and casualties in Iron Viper III except for Milans and CICVs,

b. the kills and casualties of tank destroyers from Iron Viper VI;

c. the kills and casualties of ITVs from Iron Viper VII;

d. the kills and casualties of Milans averaged over these games; and,

e. the kills and casualties of CICVs averaged over these games.

The totals produced show the individual effects of a System Study infantry battalion -- with artillery support, with tank destroyers and with ITVs -- terms of the "cost" to Blue of killing Red personnel and weapons and vehicles. I must stress that these results are completely unrealistic because they represent the effects of all these weapons when they are not acting in combination with each other. To these numbers I will compare those of Iron Viper VIII. If the synergistic increase is there, we should find that it costs Blue less in the system game than it did in the amalgamated non-system results.

PERSONNEL CASUALTIES

	RED	BLUE	COST
IV III	204	270	1.32
IV VI (TD)	20	9	0.45
IV VII (ITV)	43	2	0.05
CICV (AV)	19.0	36.3	1.91
MILAN (AV)	42.7	16.0	0.37
TOTAL	328.7	333.3	1.01
IV VIII	348	230	0.66

Personnel Casualty Comparison
Fig. 10-6

Fig. 10-6 compares personnel casualties. We see that in Iron Viper III Blue systems (remember this number does not include CICVs and Milans) killed 204 Red personnel and, in return, Blue suffered 270 personnel killed. Therefore, it "cost" Blue 1.32 personnel to kill one Red soldier -- not exactly a winning exchange rate given that Red had a six to one superiority. Likewise, taking the tank destroyer results only from Iron Viper VI we see that it "cost" Blue about half a tank destroyer crewman every time he killed a Red soldier with a tank destroyer. The rest of the numbers are worked out the same way bearing in mind that the CICV and Milan results are averaged over these three non-system games. The totals, then, for the amalgamated non-system results are that Blue killed 329 Red personnel and lost 333 in doing so. The Blue cost, therefore, was that in order to kill one Red soldier, he lost just over one. When we look at the Iron Viper VIII results, however, we see quite a change. Blue killed pretty much the same number of Red personnel (and that is a tribute to the tightness with which these games were controlled) but his cost was 100 soldiers less. In short, the result of the synergism in this particular game was that Blue was able to do the same work of destruction at two thirds the cost.

VEHICLE/WEAPON CASUALTIES

	RED	BLUE	COST
IV III	48	18	0.38
IV VI (TD)	9	8	0.89
IV VII (ITV)	18	2	0.11
CICV (AV)	3.7	28.7	7.76
MILAN (AV)	16.7	17.3	1.04
TOTAL	95.4	74.0	0.78
IV VIII	92	42	0.46

Vehicle and Weapon Casualty Comparison
Fig. 10-7

Fig. 10-7 gives the results, calculated exactly the same way, for vehicles and weapons. We see that the amalgamated non-system results are that Blue destroyed 95 Red equipments and, in doinq qo, lost 74 for a cost to destroy one Red equipment of about three quarters of a Blue equipment. Iron Viper VIII again shows about the same number of Red casualties caused but shows that Blue's cost in doing so was much less - about 60 percent less.

It looks therefore, as if the syngergistic increase of weapons in combination is substantially higher than the total effects of the weapons employed individually. Fig. 10-8 shows these results graphically from a different angle. Here we can see the Red casualties caused per hundred Blue casualties. The two bars on the left refer to personnel and the pair on the right to vehicles and weapons; in each case the bar on the left of the pair shows the amalgamated non-system results and that on the right the results from Iron Viper VIII -- the system game. In the amalgamated results, 100 Blue personnel casualties "bought" only 99 Red personnel casualties but in the system game, these same hundred casualties "bought" 152 Red kills. This is a synergistic increase of about 50 percent. In the amalgamated results 100 Blue equipment casualties "bought" 128 Red kills and in the system game the same hundred Blue casualties "bought" 217 Red kills. This is a synergistic increase of about 70 percent. These synergistic increases are neither small nor unimportant.

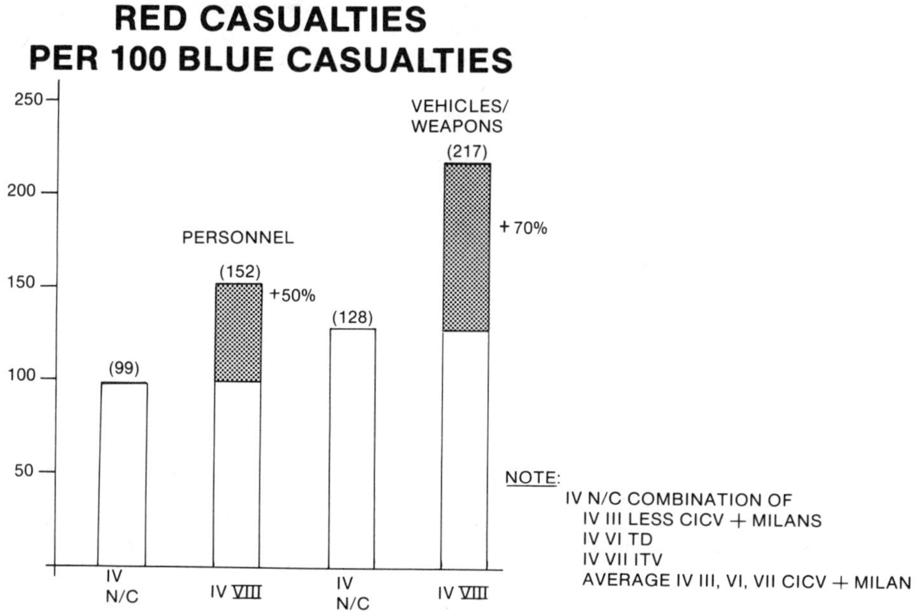

Synergistic Increases
Fig. 10-8

Before you start taking these numbers too seriously, however, I would impress upon you that they are the results of only four war games played a certain way, within certain rules, on a certain piece of ground, within certain weather conditions and so forth. The four games were not universal games producing universal truths with numbers that are certain and true for all conditions. In a real battle, or for that matter in other war games, the synergistic effect would be larger or smaller; but, these synergistic effects would be real and they would be substantial. I am not referring to mere ten per cent increases either -- I am talking about large increases of a third or more. What these numbers demonstrate, based as they are on a very carefully planned and controlled war game series, is that the synergistic increase consequent upon the proper combination of weapons is substantial and important. Indeed, I would venture that, conservatively, the synergistic increase resulting from the proper combination of weapons amounts to at least a third over the individual effects of those same weapons.

What does all this mean? Well, the first conclusion is that all that you have been taught about the benefits of using combined arms tactics is true and, in fact, even truer than you may have realized. I am sure that you all realized that TOWs get better results when they have artillery support, but I wonder how many of you realized that the overall effect will be greater than the sum of the parts. What these numbers mean is that, in all cases, weapons must be sited and used so that their effects may be combined with those of other weapons; otherwise, they will not achieve the results you expect from them. Too often one hears a statement that the effect of the TOW is such and such; yet, TOWs when properly supported by artillery and other weapons can indeed have a considerable effect -- just look at the nine for one exchange rate in the Iron Viper games -- but TOWs used alone will have very disappointing results. No weapon, used independently of others, will have much of an effect.

So, in conclusion, in military matters, one and one equals more than two.

EDITORS' NOTES

1 C. N. Donnelly, "The Wind of Change in Soviet Artillery", International Defense Review, No 6 (1982), 737-738.

2 Historical Evaluation and Research Organization, Dunn Loring, Virginia.

3 A screen is a security element whose primary task is to observe, identify, and report information; it only fights in self-defence. A guard, on the other hand, has the primary task of delaying an enemy for a period of time in addition to observing and reporting.

4 The BRDM is the standard Soviet amphibious, reconnaissance scout car and carries a crew of five. It weighs six tonnes and has four additional retractable wheels that can be used to assist traction in difficult terrain. The BRDM is also used as a launcher vehicle carrying six Sagger ATGM. The BRDM-2 is a heavier turretted version that carries three or four soldiers. John Hemsley, Soviet Troop Control (Oxford: Brassey's, 1982), 147-148.

BIOGRAPHICAL SKETCHES OF CONTRIBUTORS

DR. G. PATRICK ARMSTRONG is Head, War Game Methodology Section in the Directorate of Land Operational Research, National Defence Headquarters, Ottawa, Ontario. A graduate in philosophy from the University of Toronto, he received a Ph.D. in War Studies from King's College, University of London, in 1976.

LIEUTENANT-COLONEL IAN D. BARNES, CD, was born in Sheffield, England, in 1940. Upon graduation from high school in 1959, he was accepted into the Regular Officer Training Plan as an officer cadet in the Royal Canadian Armoured Corps. On graduation from Brandon University with a Bachelor of Science degree, he was posted to Lord Strathcona's Horse (Royal Canadians). During the period 1962 to 1969 he saw service with his Regiment in minor command and administrative positions in Canada, Cyprus, and Germany. Since 1969 he has held a variety of staff positions in NDHQ Ottawa, Headquarters Canadian Contingent Cyprus, CENTAG Headquarters (Germany), and Mobile Command Headquarters, St Hubert, Quebec. He attended staff college from 1970 to 1971 and was subsequently employed as a reconnaissance squadron commander and deputy commanding officer of his Regiment. In 1981 Lieutenant-Colonel Barnes was promoted to his present rank. He is Section Head of Combat Development Studies in the Directorate of Land Combat Development, National Defence Headquarters.

CAPTAIN JOHN CARMICHAEL received his commission as a second lieutenant of infantry in 1976 from the United States Military Academy, West Point. He was initially assigned in 1977 as a rifle platoon leader with the 3rd Mechanized Infantry Division in Schweinfurt, Germany. He subsequently served as a rifle company executive officer and an infantry battalion S3 Air in the 1st Battalion, 30th Infantry. Returning to the United States in 1980, Captain Carmichael attended the Infantry Officer Advanced Course at Fort Benning, Georgia. Since graduation in 1981, Captain Carmichael has been assigned as an offensive tactics instructor at the United States Army Infantry School.

MAJOR DAVID A. COLLETT, MC, is a serving officer in the British Army. Commissioned in 1966 from the Royal Military Academy, Sandhurst, he was posted to the 3rd Battalion, The Parachute Regiment. He subsequently served with squadrons of 21 and 22 Special Air Service (SAS) Regiments and was a helicopter pilot with the Army Air Corps. Upon return to the 3rd Battalion, The Parachute Regiment, he was appointed to command "A" Company, which he led throughout the Falkland Islands campaign. For his performance in the battle for Mount Longdon, during which action he gained first-hand experience of "the problem of the last 300 yards", Major Collett was awarded the Military Cross.

DOMINICK S. GRAHAM, MC, is currently a professor with the History Department, University of New Brunswick, and an Associate Director of the Centre for Conflict Studies. A regular officer in the Royal Artillery during the Second World War, and a major at twenty-three, he fought in four campaigns: Norway, Tobruk, The Desert, and North-West Europe. Captured in the Western Desert, he served sixteen months as a prisoner of war in Italy before successfully escaping in time to command a battery of the Guards Armoured Division for the last year of the war. He was twice wounded and mentioned in despatches and was awarded the Military Cross. Professor Graham joined the Department of History, University of New Brunswick, in 1964. He received his doctorate from London University in 1969 and assumed the chair for Military Studies at the University of New Brunswick in 1971. He was personally involved in the establishment of the Centre for Conflict Studies in 1980. Professor Graham has written a book entitled *Cassino* and has recently co-authored another, *Fire-Power: British Army Weapons and Theories of War 1904 - 1945*, with Brigadier Shelford Bidwell.

LIEUTENANT-COLONEL MICHAEL H. HANSEN received his commission in 1962 through ROTC programmes both at Texas A&M and Sam Houston State universities. His initial assignment from 1962 to 1965 was with the 8th Mechanized Infantry Division in Germany where he served as a platoon leader through company commander. On return to Fort Benning he was assigned to the 197th Infantry Brigade; for all of 1967 he was an advisor in Vietnam. Following completion of the Infantry Officer Advanced Course in 1969, Lieutenant-Colonel Hansen stayed on as an instructor at the Infantry School. Posted to Korea in 1970, he served as Secretary, General Staff, 7th Infantry Division, and as ADC to the Deputy Commanding General, 8th United States Army. Upon return to the United States, he was employed as a battalion and brigade S3 with the 4th Mechanized Infantry Division. Graduating from Command and General Staff College in 1973, he served until 1976 as an operations/plans officer in Deputy Chief of Staff Operations, Headquarters Forces Command. On return to Germany in 1976, Lieutenant-Colonel Hansen took over as G3 and, later, as Chief of Staff to the 1st Infantry Division (Forward). Rejoining the 1st

Division at Fort Riley, Kansas, in August, 1979, he commanded the 1st Battalion, 2nd Infantry, for two years. Lieutenant-Colonel Hansen has served as the Chief of Tactics, Infantry School, for the past year.

LIEUTENANT-COLONEL FRANK M. IZENOUR joined the U.S. Marine Corps in April, 1965. Upon graduation in June, 1966, from Frederick College (BA-History), Portsmouth, Virginia, he was commissioned as a second lieutenant (infantry). Following 21 weeks of training at the Basic School, Quantico, Virginia, he was assigned as a platoon commander with the 2nd Battalion, 3rd Marines, in Vietnam. Wounded for the third time, he was medevaced during May, 1967, to the United States for hospitalization. During 1968 and 1969 he was assigned to the Marine Barracks, Washington, D.C. and, in 1970, attended the Amphibious Warfare School and the Marine Advisors Course. For the next eighteen months Captain Izenour served as an advisor to the Vietnamese Marine Corps Training Center and the 1st Vietnamese Marine Battalion. Upon returning to the United States, he was assigned to the 2nd Marine Division, Camp Lejeune, N.C., where he served as a company commander for 22 months followed by an assignment as officer-in-charge of the NCO Leadership School. From 1975 to 1978 Major Izenour served as Training officer for the Marine Corps Development and Education Command (MCDEC), Quantico, Virginia. He was then transferred to Headquarters, U.S. Marine Corps, Washington, D.C., for duty in the Training Division, 1978-1979, this was followed by an assignment as Administrative Assistant to the Deputy Chief of Staff for Operations and Training. In 1981 Lieutenant-Colonel Izenour was assigned as Deputy Camp Commander, Camp Hansen, Okinawa, Japan. In 1982 he returned to the Doctrine Department, MCDEC, in the appointment of Head, Ground and Amphibious Warfare Branch.

MAJOR N. B. JEFFRIES, CD, was educated at Royal Roads Military College and the Royal Military College of Canada. On graduation in 1970, he was commissioned into the Lord Strathcona's Horse (Royal Canadians). He served on exchange with the 17/21st Lancers, British Army of The Rhine, from 1973 to 1975. Upon return to Canada in 1975 he was posted to the Regular Support Staff, Windsor Militia District, Windsor, Ontario. He returned to Regimental duty in August 1977 and subsequently saw service in both Canada and Cyprus. Following his attendance at the Canadian Forces Command and Staff College, he was posted in 1982 to his present position as Concepts Development Officer, Directorate of Land Combat Development, National Defence Headquarters, Ottawa.

MAJOR G. K. JENSEN, CD, was commissioned from the Royal Military College of Canada with a BA in History in 1965 and was appointed to the 2nd Battalion, Queen's Own Rifles of Canada, as a rifle platoon commander. In July he accompanied the 1st Battalion to Cyprus. The following year he was posted to the Combat Arms School as an instructor. From 1970 to 1973 he served with the 3rd Battalion, Princess Patricia's Canadian Light Infantry, completing his tour as a company commander. Posted subsequently to Training Command Headquarters, he returned in 1976 to the 2nd Battalion, PPCLI. Major Jensen attended Australian Army Staff College from 1980 to 1981, and on graduation was posted as Blue Room Operations Officer, Directorate of Land Operational Research, National Defence Headquarters, Ottawa, Ontario.

MAJOR LARRY LIVINGSTON, Chief of Tactics, The Basic School, enlisted in the Marine Corps in 1960 and rose to the rank of staff sergeant before being commissioned during the Vietnam War in 1968. He is an infantryman who has served in combat as both an NCO and an officer. A veteran of some of the heaviest fighting in Southeast Asia including the battles for Hue and Quang-Tri City and the heavy mechanized fighting along the DMZ during the 1972 Easter invasion by the North Vietnamese Army, he was wounded five times. He has also been awarded seventeen personal decorations, including the Navy Cross, for service in the Republic of Vietnam. The bulk of Major Livingston's active service has been spent in infantry or reconnaissance units and training commands, his billets have ranged from rifleman to executive officer of the 3rd Battalion, 5th Marine Regiment. Major Livingston is a graduate of the Marine Corps Amphibious Warfare School and the Armed Forces Staff College.

LIEUTENANT-COLONEL W. D. McKAY, CD, was commissioned from the Royal Military College in 1963 with a BA in History. Initially posted to the 2nd Battalion, Queen's Own Rifles, he was subsequently transferred to the 3rd Battalion in Germany as a platoon commander. On return of the battalion to Victoria, British Columbia, he served in several regimental duty appointments. Graduating from the one-year Canadian Land Forces Command and Staff College course in 1970, he was posted to the 1st Battalion, Princess Patricia's Canadian Light Infantry, in Calgary as a rifle company commander. From 1972 to 1975 he served with the Directorate of Land Requirements, National Defence Headquarters. In 1975 he was posted to the staff of the Director of Infantry (British Army) in Warminster, U.K. On return to Canada in 1977 he assumed the appointment of Deputy Commanding Officer, 2 Commando, and, later, Regimental Administrative Officer with the Canadian Airborne Regiment. In 1981 Lieutenant-Colonel McKay was posted to National Defence Headquarters as Head, War Game Section, Directorate of Land Operational Research.

MAJOR CHARLES R. M. MESSENGER was educated at Marlborough College, RMA Sandhurst, and Exeter College, Oxford, where he gained an honours degree in modern history, with military history as his special subject, and was awarded a Murray Naval History Prize. Commissioned into the Royal Tank Regiment in 1961, he saw service with his Regiment in the Near East, Germany, and Northern Ireland. He attended the Armor Officer Advanced Course at Fort Knox, Kentucky, and also lectured in military history there. He is a graduate of the British Army Staff College and has held staff appointments in the Ministry of Defence, including on the project management of the Combat Vehicle Reconnaissance (Wheeled) and (Tracked) projects. Although he retired from the Regular Army in 1980 in order to pursue a full time career as a writer in military matters, Major Messenger is still actively serving in the Territorial Army and works as a part-time staff officer with an armoured division headquarters in Germany. Among his published works are Trench Fighting 1914 - 1918 (Ballantine, 1972) and The Blitzkrieg Story (Scribner, 1976, published in the UK by Ian Allan as The Art of Blitzkrieg). He has recently completed studies on three crucial turning points of World War II -- First Alamein, the first thousand bomber raid, and the critical convoy battle of the Battle of the Atlantic -- as well as a book on the Tunisian Campaign, which will be published in North America by Hippocrene. He is currently holding a research fellowship at the Royal United Services Institute in London, studying military aircraft procurement, and writing a number of books, among them a study of the NATO concept of the anti-armour battle, an appraisal of "Bomber" Harris and the strategic bombing campaign of 1939 - 1945, and the latest volume of the history of the 6th Gurkha Rifles and 2nd Battalion Scots Guards in the Falklands Campaign. He also contributes articles and book reviews to a number of publications.

LIEUTENANT-COLONEL BARRY MONTGOMERY, a native of Starkville, Mississippi, graduated from Mississippi State University in 1964. Commissioned a second lieutenant from the Armor Officer School, Fort Knox, Kentucky, he served from 1964 to 1966 with the 2nd Tank Battalion, Force Troops, Camp Lejeune, N.C., as both a platoon leader and executive officer of a tank company. From 1966 to 1968 he was assigned as the executive officer, Marine Barracks, NAS Sanford, Florida. In 1968 he was transferred to the 1st Tank Battalion, 1st Marine Division, Republic of Vietnam, where he served as a tank company commander and S-3 operations officer. Upon returning from Vietnam, he was assigned once again as a tank company commander to the 2nd Tank Battalion, Camp Lejeune, N.C. In 1971 he was assigned to the Armor Officer Advanced Course, Fort Knox, and, on graduation in 1972, was posted to the Schools Battalion, Marine Corps Base, Camp Pendleton, California. In 1976 he attended the Marine Corps Command and Staff College and, on graduation, was assigned to 3rd Marine Division, Okinawa, Japan, as division adjutant. Upon return to the United States in 1978, he was assigned to the Development Center, Quantico, Virginia and has served as the project officer for tanks, the mobile protected weapons system, and the light armored vehicle programme.

MAJOR G. H. SAWATSKY, CD, is an artillery officer who began service with the 2nd Regiment, Royal Canadian Horse Artillery (2 RCHA) in 1956. In 1960, following training at the Royal Canadian School of Infantry and the Royal Canadian School of Artillery, he was posted to 3 RCHA and served in Germany and Winnipeg. He returned to the School of Artillery as a student and as an instructor-in-gunnery (IG) until 1960. He completed the German Command and Staff Course in 1970 and was posted to Canadian Forces Base Shilo where he held several staff positions. From 1977 to 1981 he served in the Directorate of Training, National Defence Headquarters (NDHQ), Ottawa. He is currently on staff of the Directorate of Land Operational Research as the Blue Room Commander.

BRIGADIER (RETD) RICHARD E. SIMPKIN left Winchester College to go to Trinity College, Cambridge, just as war broke out. After one year, in which he switched from medicine to engineering, he joined the Royal Tank Regiment, where he remained for 30 years. A period of active service in the Middle East, during which he was awarded the Military Cross, ended when he was taken prisoner in June, 1942. It was at this point, with the very strong motivation of escape, that his interest first turned to modern languages. Since the war his service has taken him to most parts of the world, including Hong Kong and other areas of the Far East. From 1946 to 1948 he spent an interesting period on technical, industrial and economic intelligence. Graduating from Staff College in 1951, he went straight on to the Royal Military College of Science, graduating from the Technical Staff Course with a specialisation in vehicles in 1953. An instructional tour in 1957-59 was divided between the directing staffs of the Staff College and the Royal Military College of Science (Combat Vehicle Technology). From 1960 to 1963 he headed the equipment branch of the Royal Armoured Corps Directorate. In 1963 he received the award of Officer of the Order of the British Empire (OBE) and took command of 1st Royal Tank Regiment; most of his time in command was spent in British Army of the Rhine, both in an infantry brigade and as battle group commander in an armoured brigade. For a few months before handing over, he took the Regiment to Aden in a mainly dismounted counter-insurgency role but with an armoured squadron held embarked, based on Bahrain. He then had a short second tour at the Royal Military College of Science as Military Director of Studies (Weapons and Vehicles). Promoted Brigadier in 1968, he was appointed Director of Operational Requirements 3 (Army) in the Ministry of Defence, covering equipment policy for the direct fire battle and all aspects of mobility. In 1971 he decided to retire early to give himself time to develop a second career. Fluent in German, French, and Russian, he established a language consultancy which he expanded into a private company in 1973. His clientele currently includes a number of British national agencies but is mainly drawn from France, Germany and the USA. He now divides his time between authorship, consultancy and lecturing, specialised commercial translation, and book translations/multilingual editing. Among his military works are Tank Warfare (1979),

Mechanized Infantry (1980), Anti-Tank (1982), Human Factors in Mechanized Warfare (1983) and Red Armour (1984). Brigadier Simpkin has also published numerous essays and articles in several military publications.

PETER H.VIGOR is Professor Emeritus with the Soviet Studies Research Centre, of which he is founder and original Director, at the Royal Military Academy, Sandhurst. He is a graduate of the University of Southampton and Cambridge. In the Second World War he was sent to help get the Polish army out of the USSR and to organize and train what was subsequently to become II Polish Corps under General Anders. Fluent in both Russian and Polish, Professor Vigor has written numerous authoritative articles on Soviet military affairs. He is also the author of several books, among them The Soviet View of War, Peace and Neutrality and A Guide to Marxism and Its Effects on Soviet Development. His latest work, Soviet Blitzkrieg Theory, is the first Western attempt to look at, evaluate, and understand Russian thinking on this crucial issue. Based almost entirely on Soviet sources, it deals realistically and knowledgeably with the evolution of their theories and the practical uses to which they may be put.

LIST OF ABBREVIATIONS

AA - anti-aircraft
AD - air defence
adm - administration
admin - administration (US)
adv - advance
AEV - armoured engineer vehicle
APDS - armour piercing discarding sabot
AFV - armoured fighting vehicle
AG - assault gun
AMV - armoured maintenance vehicle
AP - armour piercing
APC - armoured personnel carrier
armd - armoured
arty - artillery
ARV - armoured recovery vehicle
ASL - assistant squad leader
aslt - assault
AT - anti-tank
ATGM - anti-tank guided missile
ATGW - anti-tank guided weapon
Atk - anti-tank
av - average
AVLB - armoured vehicle launched bridge
AVR - armoured vehicle reconnaissance
BAR - Browning automatic rifle (US)
BC - battery commander
BCFV - Bradley cavalry fighting vehicle
BIFV - Bradley infantry fighting vehicle
bn - battalion
BTR - brone transporter (Soviet armoured personnel carrier)
cas - casualty(ies)
CB - counter-battery
CFV - cavalry fighting vehicle
CM - counter-measures, or counter-mortar
Co - company (US)
comd - commander
comms - communications
comp - composite
conc - concentration
coy - company
Cpl - corporal
CSV - combat support vehicle
CTC - Combat Training Centre
DAG - divisional artillery group
def - defence
det - detachment
dir - direction
div - division
DLCD - Directorate of Land Combat Development
DLOR - Directorate of Land Operational Research
DS - direct support
DComd - deputy commander
ech - echelon
en - enemy
engr - engineer

fd - field
FEBA - Forward Edge of the Battle Area
FFR - free flight rocket
FGA - fighter ground attack
FOO - forward observation officer
FPW - firing-point weapon
FUP - forming up place
GD - general duties
gd(s) - guard(s)
gp - group
GPMG - general purpose machine gun
GPMG(SF) - general purpose machine gun (sustained fire)
GSFG - Group of Soviet Forces Germany
H&S - headquarters and service
HAW - heavy anti-tank weapon
HEDP - high explosive dual purpose
HEIT - high explosive incendiary tracer
hel - helicopter
HMG - heavy machine gun
HQ - headquarters
hy - heavy
ICM - improved conventional munitions
IFV - infantry fighting vehicle
IGF - Inner German Frontier
IMAAWS - infantry man-portable anti-armour/assault weapon system
instr - instructor
ITV - improved TOW vehicle
IWS - individual weapon sight
km - kilometre
LAV - light armoured vehicle
LAW - light anti-tank weapon
LOB - left out of battle
lt - light
LCpl - lance corporal
MAB - Marine amphibious brigade
MAF - Marine amphibious force
mag - magazine
maint - maintenance
MAU - Marine amphibious unit
MAW - medium anti-tank weapon
MBRL - multi-barrel rocket launcher
MBT - main battle tank
mech - mechanized
med - medium or medical
MG - machine gun
MICV - mechanized infantry combat vehicle
mor - mortar
MR - motor rifle
MRR - motor rifle regiment
MT - motor transport
M/C - motorcycle
NBC - nuclear, biological, and chemical
NCO - non-commissioned officer
NOE - nap of the earth
obj(s) - objective(s)
OMG - operational manoeuvre group
OP - observation post

op - operation, operational
ORBAT - order of battle
PGM - precision guided munitions
PL - platoon leader (US)
pl - platoon (Canadian)
plat - platoon (US)
PNGs - passive night goggles
posn - position
prep - preparation
Pte - private (Canadian)
ptl - patrol
Pvt - private (US)
RCL - recoiless rifle (anti-tank)
rds - rounds
recce - reconnaissance
regt - regiment
res - reserve
RL - rocket launcher
RMA - Royal Military Academy
SAW - squad automatic weapon
SC - scout car
sec - section
sect - section
SFF - self-forging fragment
Sgt - sergeant
sigs - Signals
SL - squad leader (US)
SL - start line
SLAP - saboted light armour penetrator
SLR - self-loading rifle
SMG - sub-machine gun
SOP - standing operating procedures
SP - self-propelled
sp - support
sqd - squad
sqn - squadron
SSgt - staff sergeant
svc - service
tac - tactical
TD - tank destroyer
tech - technical
TF - task force
tk - tank
tm - team
TOW - tube-launched, optically-tracked, wire-guided ATGM
tpt - transport
USATACOM - United States Army Tank and Automotive Command
van - vanguard
veh - vehicle
V/STOL - vertical/short take-off and landing
WP - white phosphorous

BIBLIOGRAPHY AND SUGGESTED READING

Antal, Captain John F. "Unit Battle Drills", Infantry, No 4 (July - August, 1983), 31-32.

"Battle Drill", Infantry School Quarterly, (January, 1957), 6-16.

Baxter, Lieutenant-Colonel William P. "Soviet Communications: Bare-Bones but Secure", Army, No 8 (August, 1982), 30-33.

Beeler, Lieutenant-Colonel R.A. "New Infantry Battalion Structure: An Overview", Marine Corps Gazette, No 6 (June, 1983), 17-19.

Besch, Captain Edwin W. "Mechanizing the Marines", Marine Corps Gazette, No 2 (February, 1983), 17-19.

______________________. "Infantry Fighting Vehicles: Their Evolution and Significance", Marine Corps Gazette, No 7 (July, 1983), 50-60.

______________________. "IFVs - A Matter of Choices", Marine Corps Gazette, No 1 (January, 1984), 26.

Bidwell, Shelford, and Dominick Graham. Fire-Power. London: George Allen and Unwin, 1982.

Biryukov, Major-General G., and Colonel G. Melnikov. Antitank Warfare. trans. David Myshne. Moscow: Progress Press, 1972.

Bonwit, Captain David F. "Who Coordinates Fire Support?", Marine Corps Gazette, No 9 (September, 1983), 25-26.

Bradley, Clifford D. "Wheels Versus Tracks", Armor. (May - June, 1981), 24-29.

Burke, Major J.D. "Maneuver Warfare and the MAGTF", Marine Corps Gazette, No 9 (September, 1982), 67-71.

Canby, Steven L. Classic Light Infantry and New Technology. Arlington: Defense Advanced Research Project Agency, 1983 (Contract No. MDA 903-81-C-0207).

Caravelli, John M. Soviet and Joint Warsaw Pact Exercises: Functions and Utility", Armed Forces and Society, No 3 (Spring, 1983), 393-426.

Carter, Captain R.G. "Mechanization: United or Divided", Marine Corps Gazette, No 8 (August, 1982), 26-27.

Carver, Field Marshal Lord. The Apostles of Mobility. London: Weidenfeld and Nicolson, 1979.

Chaudrue, Lieutenant-Colonel Robert G. "Requiem For The Infantry", Infantry, No 3 (May - June, 1978), 28-31.

Chisum, Captain M.C. "Combined Arms Properly Understood", Marine Corps Gazette, No 11 (November, 1982), 27-30.

Clifford, Lieutenant-Colonel Kenneth J. Progress and Purpose: A Developmental History of the United States Marine Corps 1900 - 1970. Washington: History and Museums Division, Headquarters, USMC, 1973.

Conway, Captain Paul L. "The Assault", Infantry, No 6 (November - December, 1982), 19-23.

Cooper, Bryon. The Ironclads of Cambrai. London: Souvenir Press, 1967.

DePuy, General William E. "One-Up and Two-Back?", Army, No 1 (January, 1980), 20-25.

Despres, John, Lilita Dzirkals, and Barton Whaley. Timely Lessons of History: The Manchurian Model for Soviet Strategy. Report prepared for the office of the Secretary of Defense/Director of Net Assessment. Santa Monica: Rand, 1976.

Dick, C.J. "Soviet Operational Manoeuvre Group", International Defense Review, No 6 (1983), 770-776.

Donnelly, C.N. "The Development of Soviet Military Doctrine", International Defense Review, No 12 (1981), 1589-1596.

______________. "The Wind of Change in Soviet Artillery", International Defense Review, No 6 (1982), 737-744.

______________. "The Soviet Operational Manoeuvre Group", International Defense Review, No 9 (1982), 1177-1186.

English, John A. A Perspective on Infantry. New York: Praeger, 1981.

Estes, Major Kenneth W. "More on Mechanizing Marines", Marine Corps Gazette, No 7 (July, 1983), 58-59.

Fuller, Major-General J.F.C. Lectures on F.S.R. II. London: Sifton Praed, 1931.

___________________________. The Army in My Time. London: Rich and Cowan, 1935.

___________________________. "Armour and Counter-Armour", Infantry Journal (March - April - May, 1944), 19-24, 42-48, and 39-43.

___________________________. Thunderbolts. London: Skeffington and Son, 1946.

___________________________. Decisive Battles of the Western World. ed. John Terraine. London: Granada, 1981.

Gale, General Sir Richard N. "Infantry in Modern Battle", Canadian Army Journal, No 1 (January, 1955), 52-61.

Gritz, Major John P. "Light Armoured Vehicles or Light Armoured Victims", Marine Corps Gazette, No 8 (August, 1982), 36-42.

Gritz, Major John P. "Dragons in the Infantry Squad", _Marine Corps Gazette_, No 9 (September, 1982), 26-28.

Guizonnier, Lieutenant-Colonel C. _Anti-Tank Training in the French Army_. Unpublished.

Hamilton, Major C.F. "LAV and the Amphibious Assault", _Marine Corps Gazette_, No 10 (October, 1983), 66-73.

Hastings, Max and Simon Jenkins. _The Battle for the Falklands_. New York: W.W. Norton, 1983.

Hemsley, John. _Soviet Troop Control_. Oxford: Brassey's, 1982.

Hewish, Mark. "FFV's Lightweight AT4", _International Defense Review_, No 5 (1982), 612-614.

Hogg, Ian V. "MICVs, IFVs and APCs", _Defence_, No 8 (August, 1982), 443-447.

Insight Team of the Sunday Times. _Insight on the Middle East War_. London: Deutsch, 1974.

____________________. _War in The Falklands_. New York: Harper and Now, 1982.

Jeapes, Lieutenant-Colonel A.J. "Letter to the Editor", _British Army Review_, No 63 (December, 1979), 73-74.

Jenkins, D.H.C. "T-34 to T-80", _International Defense Review_, No 12 (1981), 1649-1654.

__________. "Abrams and Leopard 2", _International Defense Review_, No 12 (1981), 1657-1684.

__________. "Battle Tank Design: Some Thoughts for the Future", _International Defense Review_, No 2 (1982), 171-176.

__________. "Operating the Bradley Infantry Fighting Vehicle", _International Defense Review_, No 9 (1982), 1203-1206.

King, Captain W.A. "Rifle Squad Remains Key Element", _Marine Corps Gazette_, No 6 (June, 1983), 26-30.

Kippenberger, Major-General Sir Howard. _Infantry Brigadier_. London: Oxford University Press, 1949.

Leader, Captain C.A. "Maneuver Tactics and The Art of War", _Marine Corps Gazette_, No 3 (March, 1983), 52-57.

Liddell Hart, B.H. _Sherman_. London: Ernest Benn, 1930.

McCaslin, Major James K., Jr. "The BMP in Combat", _Infantry_, (January - February, 1977), 27-31.

McDermott, Captain David F. and Captain Scott R. Gourley. "Evolution of the BMP", _Infantry_, No 6 (November - December, 1983), 19-22.

Mellenthin, Major-General F.W. von. Panzer Battles. trans. H. Betzler. Norman: Oklahoma University Press, 1983.

Messenger, Charles. The Art of Blitzkrieg. London: Ian Allan, 1976.

__________________. "Current Thoughts in Armour", Defence, No 6 (June, 1982), 318-322.

Miksche, Major F.O. Blitzkrieg. London: Faber and Faber, 1941.

Miller, D.M.O. "The Infantry Combat Vehicle: An Assesqment", Military Technology and Economics, Issue 9 (May - June, 1979), 29-38.

Minor Lessons Learnt in the Falklands. Aldershot: 3 Para After-Action Report, 1983.

Moore, Captain R.S. "Two Teams For Rifle Squad", Marine Corps Gazette, No 9 (September, 1983), 20-22.

Morgan, General Sir William. "The Revival of Battle-Drill in World War 2", Army Quarterly and Defence Journal, No 1 (October, 1973), 57-60.

Murray, Major Terrence P. "Establishing Force Structure and Doctrine", Marine Corps Gazette, No 9 (September, 1983), 60-63.

NATOs Fifteen Nations. Special Issue 2 (1981).

Neller, Captain R.B. "LAV: What it Can Do", Marine Corps Gazette, No 10 (October, 1983), 72-73.

Ogorkiewicz, R.M. "Brandt Gun-Mortars", International Defense Review, No 2 (1982), 201-204.

__________________. "MCV 80: The New British Infantry Combat Vehicle", International Defense Review, No 6 (1982), 721-726.

Ozolek, Major David J. "Squad Training", Infantry, No 6 (November - December, 1982), 16-18.

Pappa, 1st Lt L.M. "More Marines for Mortar Section", Marine Corps Gazette, No 9 (September, 1983), 22-24.

Patton, General George S. Jr. War As I Knew It. New York: Pyramid, 1970.

Peele, Captain R.B. "New Mortar More Accurate, Lethal", Marine Corps Gazette, No 6 (June, 1983), 30-31.

Philipp, Udo. "NATO Strategy Under Discussion in Bonn", International Defense Review, No 9 (1980), 1367-1371.

Post Operational Report. Aldershot: 3rd Battalion, The Parachute Regiment, 1982.

Post Operational Report of the Second Battalion, The Parachute Regiment.

"Post-War Development of Anti-tank Tactics", Canadian Army Journal, (December, 1952), 4-14.

Rilhac, Colonel Andre L. "Commando Training for French Armor", Armor (September - October, 1983), 18-20.

Rogers, Captain J.P. Jr. "Don't Cut the Rifle Squad", Marine Corps Gazette, No 8 (August, 1982), 31-32.

Rossetto, Captain L. "Modern Soviet Infantry Weapons", Canadian Defence Quarterly, No 1 (Summer, 1982), 18-21.

Ruggero, Lieutenant Edward J. "Digging In: Platoon Defense", Infantry, No 1 (January - February, 1983), 20-23.

Schmitt, 2nd Lt John F. "Heavy MG Section in Crawl Stage", Marine Corps Gazette, No 6 (June, 1983), 21-23.

________. "Improving Our Antiarmour Punch", Marine Corps Gazette, No 6 (June, 1983), 24-26.

Scott, Harriet Fast, and William F. Scott. The Armed Forces of the USSR. Boulder: Westview Press, 1979.

________. The Soviet Art of War. Boulder: Westview Press, 1982.

Senger und Etterlin, General Frido von. Neither Fear Nor Hope. London: Macdonalds, 1963.

Simpkin, Richard E. Tank Warfare. London: Brassey's, 1979.

________. Mechanized Infantry. Oxford: Brassey's, 1980.

________. "No Speed Please, We're British", Jane's 1981-82 Military Annual. Ed. Colonel John Weeks. London: Jane's, 1981.

________. "Closing the Survivability Gap", Armor, No 6 (November - December, 1981), 19-24.

________. Antitank. Oxford: Brassey's, 1982.

________. "The Roles of Infantry in the Manoeuvre Battle", Jane's Military Review. Ed. Ian V. Hogg. London: Janes', 1982.

________. "Hammer, Anvil and Net", British Army Review, No 72 (December, 1982), 11-21.

________. "The Renaissance of the Light Combat Vehicle", NATO's Sixteen Nations. (June - July, 1983), 89-96.

________. "When the Squad Dismounts", Infantry, No 6 (November - December, 1983), 14-18.

Smith, Robert C. "Tank Killers", Infantry, (January - February, 1981), 8-11.

Smith, Captain W.L. "Major Impacts on Weapons Company", Marine Corps Gazette, No 6 (June, 1983), 19-20.

Snelson, Major Robert L. "The New Infantry Battalion Structure", Marine Corps Gazette, No 7 (July, 1982), 25-26.

Stolfi, Russel H.S., Lonnie O. Ratley, III, and John F. O'Neill, Jr. German Disruption of Soviet Command, Control and Communications in Babbarossa, 1941. Washington: Director, Net Assessment, Office of the Secretary of Defense, 1982.

Suvorov, Viktor. The Liberators. London: Hamish Hamilton, 1981.

_______________. Inside the Soviet Army. New York: Macmillan, 1982.

Szelowski, 1st Lt D.W. "Using Dragons Effectively", Marine Corps Gazette, No 9 (September, 1982), 28-29.

The BDM Corporation. Generals Balck and von Mellenthin on Tactics: Implications for NATO Military Doctrine. (19 December, 1980) Director of Net Assessment, Office of the Secretary of Defense and the Defense Nuclear Agency, Contract No. DNA 001-78-C-0114.

"The Employment of Non-Mechanized Infantry", Report of a Symposium Jointly Sponsored by the RUSI and the Commander, ACE Mobile Force (Land) held in Hamburg on 28 April, 1980.

Vigor, P. H. "Soviet Army Wave Attack Philosophy: The Single-Echelon Option", International Defense Review, No 1 (1979), 43-46.

Vorobyov, Colonel I. "Weapons and Tactics", Soviet Military Review, No 4 (April, 1977), 12-14.

Weeks, Colonel John. "The Modern Infantryman", Military Technology and Economics, Issue 9 (May - June, 1979), 23-26.

Whittle, 1st Lt Frederick J. "13 to 11: What's so Unsound?", Marine Corps Gazette, No 11 (November, 1982), 44-45.

Wilmot, Chester. The Struggle for Europe. London: Collins, 1952.

Wilson, Major G.I. and Major W.A. Woods. "Defending Maneuver Concepts", Marine Corps Gazette, No 8 (August, 1982), 34.

Wyly, Lieutenant-Colonel Michael D. "Thinking Beyond the Beachhead", Marine Corps Gazette, No 1 (January 1983), 34-38.

__________________________. "War Without Firepower?", Marine Corps Gazette, No 3 (March, 1983), 17-18.

Woollcombe, Robert. The First Tank Battle: Cambrai 1917. London: Arthur Barker, 1967.